Cambridge Studies in Ecology presents balanced, comprehensive, up-to-date, and critical reviews of selected topics within ecology, both botanical and zoological. The Series is aimed at advanced final-year undergraduates, graduate students, researchers, and university teachers, as well as ecologists in industry and government research.

It encompasses a wide range of approaches and spatial, temporal, and taxonomic scales in ecology, including quantitative, theoretical, population, community, ecosystem, historical, experimental, behavioral and evolutionary studies. The emphasis throughout is on ecology related to the real world of plants and animals in the field rather than on purely theoretical abstractions and mathematical models. Some books in the Series attempt to challenge existing ecological paradigms and present new concepts, empirical or theoretical models, and testable hypotheses. Others attempt to explore new approaches and present syntheses on topics of considerable importance ecologically which cut across the conventional but artificial boundaries within the science of ecology.

T0296334

Fire and vegetation dynamics:
Studies from the North American boreal forest

CAMBRIDGE STUDIES IN ECOLOGY

Fire and vegetation dynamics:

Studies from the North American boreal forest

EDWARD A. JOHNSON

Division of Ecology, Department of Biological Sciences, University of Calgary

CAMBRIDGE
UNIVERSITY PRESS

Published by the Press Syndicate of the University of Cambridge
The Pitt Building, Trumpington Street, Cambridge CB2 1RP
40 West 20th Street, New York, NY 10011-4211, USA
10 Stamford Road, Oakleigh, Melbourne 3166, Australia

First published 1992
Reprinted 1995

A catalogue record for this book is available from the British Library

Library of Congress cataloguing in publication data

Johnson, Edward A. (Edward Arnold), 1940-
    Fire and vegetation dynamics: studies from the North American
boreal forest / Edward A. Johnson.
        p.   cm. – (Cambridge studies in ecology)
    Includes bibliographical references and index.
    1. Taiga ecology – North America.   2. Forest fires – Environmental
aspects – North America   3. Fire ecology – North America.
4. Vegetation dynamics – North America.   5. Trees – North America –
Ecology.   I. Title.   II. Series.
QK110.J64   1992
581.5'2642-dc20   91-36693   CIP

ISBN 0 521 34151 5 hardback
ISBN 0 521 34943 5 paperback

Transferred to digital printing 2004

SE

Contents

Preface

Two solitudes: fire behavior and fire effects

This book attempts to couple four characteristics of fire behavior in the boreal forest to their effects on boreal tree populations. The presentation is *unbalanced* with the discussions of fire behavior more detailed than the discussions of fire effects on the plants. This is because our knowledge of fire behavior (intensity, rate of spread, duff consumed, frequency of occurrence, etc.) has developed by a quantitative understanding of the physical processes of fire while our understanding of fire effects on populations has been largely descriptive. As ecologists, we have shown a surprising lack of curiosity about how the fires actually produced their ecological effects (Van Wagner and Methven 1978). This is a result of a strong phytosociological tradition in plant ecology in which an adequate explanation was description of species composition patterns and correlation with general environmental factors. No attempts were made to specify the causal connections between fire behavior and individual plants in terms of appropriate physical variables. Nor were the fire effects on individuals tied to population recruitment and mortality processes.

This book should be of interest to those who already have some knowledge of populations and community ecology but wish an introduction to fire behavior and how it might be coupled to population processes. I have made no attempt to review all of the fire ecology literature of the boreal forest, nor do I discuss all aspects of fire in the boreal forest. Instead, I have concentrated on studies which have coupled the specific physical understanding of fire behavior to individual plants and populations. The book might be seen as a demonstration of the need for fire effects research which takes this process–response approach. To the best of my knowledge, only fire ecology has a well-developed understanding of the physical disturbance processes in the terrestrial environment although a process–response approach is developing in windthrow (Deans and Ford 1983, Petty and Swain 1985, Schaetzl *et al.* 1989, Blackwell *et al.* 1990). In the marine intertidal, Mark Denny's 1988 book *Biology and the Mechanics of the Wave-Swept Environment* stands as a classic of the process–response approach.

The book is divided into three parts. The first part (Chapter 1) gives the basic thesis of the book: that major aspects of the vegetation dynamics of the North American boreal forest can be understood by quantifiable aspects of forest fire behavior. The second part of the book deals with fire and how it is related to vegetation. Chapter 2 argues that the position of the Arctic airstreams sets the seasonal and geographic extent of the fire season in the boreal forest and that specific synoptic scale weather patterns are responsible for the ignition and spread of large fires.

The next four chapters give the relationships between specific physical processes of fire behavior and the population processes of recruitment and mortality. Chapter 3 discusses fire spread and its dependence on fuel, weather and topography with implications for seed dispersal and mortality. Chapter 4 considers the heat output from the flaming fire front, and its effect on plant mortality. Chapter 5 deals with the consumption of duff by the fire and its effect on recruitment and mortality. Chapter 6 discusses fire recurrence and its effect on survivorship and forest age patterns. Finally, the last part of the book (Chapter 7) combines fire behavior and an understanding of its effects on population recruitment and mortality to explain the observed tree age distributions in the boreal forest.

The book will consider mostly tree populations since more is known about their dynamics and coupling to fire behavior than herbs and shrubs. I have limited myself to primary sources and ideas supported by empirical data. I have tried not to use hypothetical diagrams which do not have explicit empirical tests. Further, I have restricted myself almost completely to fire behavior and effects studies in the boreal forest (see Figure 1.1). By doing so, I hope to minimize the danger of using studies which are not appropriate or comparable to the boreal forest. In the past, we have all used research and arguments from other vegetation types and fire regimes to bolster our interpretations when we lacked our own empirical data. Without confirmatory studies, this approach has led to serious confusion. I will also try to develop a quantitative understanding of fires and populations since both fires and populations are not easy subjects to develop intuitive feelings about and the development of equations often exposes the level of sophistication in our understanding of the processes.

The chemical aspects of combustion do not play a central role in our discussion of forest fires because of the small variation in composition of woody and herbaceous matter. However, the availability of these fuels (generally determined by their drying rates) and heat and mass transfer rates associated with fire do play a central role. If we were interested in nutrient cycling instead of population dynamics, our focus would of necessity include more of the chemistry of combustion and resulting soil and plant chemical changes.

Acknowledgements

I should like to thank the following persons for reading the entire book and making helpful and clarifying comments: Marty Alexander, John Birks, Yves Bergeron, Gina Fryer, David Greene and Paul Zedler. Also many people provided specific comments, information, references and preprints: Bill Archibold, Z. Chrosciewicz, Joan Foote, Louise Filion, David Foster, George La Roi, Hank Lewis, Vic Lieffers, Craig Lorimer, Serge Payette, Brian Stock, Ross Wein, Dana Wowchuk, Keith Van Cleve.

A reading of this book will indicate the importance to me and the field of fire behavior and effects of Charlie Van Wagner. Without his outstanding research over the last 30 years, we would all be significantly poorer.

The manuscript was carefully and expertly typed innumerable times by Eileen Muench. The editorial assistance of Maria Murphy and Jane Bulleid was most helpful.

Finally, I should like to dedicate this book to my daughter Joanne who has helped me in the field since she could count and write.

Tree species mentioned in the text

English name	French name	Botanical name
Ground juniper	Genévrier commun	*Juniperus communis* L.
Balsam fir	Sapin baumier	*Abies balsamea* (L.) Mill.
White spruce	Epinette blanche	*Picea glauca* (Moench) Voss
Black spruce	Epinette noire	*Picea mariana* (Mill.) BSP
Jack pine	Pin gris	*Pinus banksiana* Lamb.
Red pine	Pin rouge	*Pinus resinosa* Ait.
White pine	Pin blanc	*Pinus strobus* L.
Speckled alder	Aulne rugueux	*Alnus rugosa* (Du Roi) Spreng.
White birch	Bouleau à papier	*Betula papyrifera* Marsh.
Trembling aspen	Peuplier faux-tremble	*Populus tremuloides* Michx.

1

Fire and the boreal forest: the process and the response

Fire behavior in the boreal forest (Figure 1.1) has at least four characteristics which are important in understanding the dynamics of its populations: crown fires, the size of the area burnt, the frequency with which areas burn, and the amount of the forest floor which is ashed.

More precisely, fires in the boreal forest have high frontal intensity (heat output in $kW\ m^{-1}$) or flame length at the flaming front. The high intensity results in a crown fire regime shared by only a few other North American ecosystems e.g. grasslands, chaparral and montane conifer forests. Some of the largest fires in the world have been reported in the boreal forest. Murphy and Tymstra (1986) give the size of a 1950 free-burning fire in northern British Columbia and Alberta as 1.4 million hectares. Fires greater than 100 000 ha are common with the interval between fires being on average about every 100 years. The forest floor of upland boreal forest has a depth, moisture content and bulk density which makes it very flammable so that large amounts of mineral soil are exposed.

These four fire behavior characteristics affect the population processes of recruitment and death. The high frontal fire intensity causes crown scorch and cambial death which results in both canopy and understory tree mortality. Rapid rates of fire spread lead to large areas burnt and greater dispersal distances for trees that do not have serotinous cones and must seed-in from outside the burn or from unburned patches within the burn. The frequency of fire is about half the trees' non-fire life span. The mineral soil exposed by the fire gives higher recruitment for most boreal trees than does a duff-covered surface. Buried viable seeds are not an important regeneration strategy after disturbance because of the large amount of duff consumed.

The tree population dynamics reflect the mortality and recruitment forces of fire. The age class distributions have a cohort structure in which

1

the understory does not generally replace the canopy. This results in a time-lag in canopy replacement, the lag being related to the fire frequency, fire intensity and duff removed. Consequently, the local population dynamics over several generations appear to fluctuate at least in part as a result of the stochasticity of the fire behavior.

The objective of this book is to examine the coupling between fire behavior and populations. In the first part of this chapter, I gave a very general blueprint of the process–response approach which will be used. In the remaining chapters I will attempt to construct this coupling more precisely. The conceptual framework involves coupling the fire to the biophysics of the organism, and this in turn to the population dynamics (cf. Johnson 1985). In order to do this, concepts of heat transfer and combustion of forest fires must be understood. Relevant variables, coefficients and parameters of forest fire ignition and spread can then be chosen which cause biological effects in the individual organisms. This in turn can be related in a direct manner to the recruitment, mortality and age class distributions of the populations.

Hopefully, the readers will find many things I say unsatisfactory so that they will be driven into the field and laboratory to correct these confusions and misunderstandings. I have tried to resist the temptation to speculate or give hypothetical arguments based on casual observations.

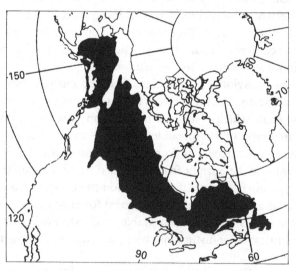

Figure 1.1. The boreal forest in North America (from Hare and Ritchie 1972).

2

Fires and climate

Weather and climate control fire occurrence and spread in the boreal forest. As we shall see in this chapter, airstream movements and boundaries limit the length of the fire season and determine the seasonal geographic progression of fires. Further, a critical synoptic weather pattern of upper level (50 kPa) ridge build-up and breakdown is often responsible for the ignition and spread of large and widespread fires. The weather determines these patterns by generating the principle ignition source (lightning), controlling the most significant variable of fuel flammability (fuel moisture) and providing the high winds required for rapid spread.

Fire occurrence

The occurrence of forest fires in the boreal forest of Canada is given in Figure 2.1. This map was constructed from over 40 000 individual forest fires from all causes for 1961–6. Although the period is short, and not all areas are covered with the same thoroughness (some areas, e.g. northern Quebec, have no records at all), the map still presents the characteristic pattern of fires in the boreal forest (Simard 1975). Because of the scale of the map, the localized high fire occurrence in Figure 2.1 associated with population centers (e.g. La Ronge, Saskatchewan and Chicoutimi, Quebec) and transportation corridors (e.g. south from Moosonee, Ontario) have been removed. Also less obvious is the higher fire occurrence in the summer cabin region running from north of Toronto to north of Montréal and the large area of reduced fire occurrence southeast of North Bay, Ontario caused by Algonquin Provincial Park. However, when these man-caused effects on the fire occurrence are removed, a clear decrease in fire occurrence can be seen from the southern to the northern boundary of the boreal forest and from west to east.

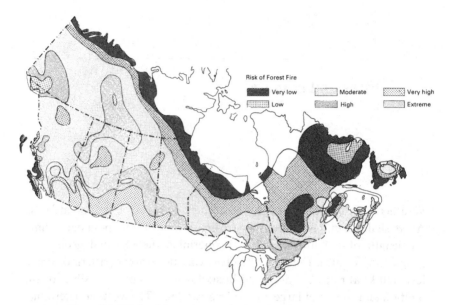

Figure 2.1. Wildland fire occurrence in Canada (cf. Simard 1975).

Indian-caused fires

The effect of indigenous peoples on fire occurrence in the boreal forest before European influence can only be speculated on at this time. However, any argument on Indian-caused fires must be based on an understanding of the Indians' life ways and how these were changed by the arrival of Europeans. Since ecologists are not usually familiar with Indian life ways, I will spend a few pages outlining them and their significance in man-caused set fires.

European arrival did not destroy the boreal forest Indians' culture as thoroughly as it did the eastern deciduous forest and plains Indians' culture. This was because in the boreal forest there was no serious competition between Europeans and Indians for land or natural resources. In fact, both the French-Canadian traders from Montréal and the Hudson Bay Company were interested in having the Indians devote themselves completely to trapping and discouraged or actively opposed any European activities which might have turned the Indians away from trapping (Heidenrich and Ray 1976).

The Indians' seasonal cycle of activities appears to have remained largely unchanged from at least the 1600s until about 1945. Most boreal forest

Indians lived in small, multiple family hunting groups which were organized into loose regional bands. They seasonally moved to areas of resource concentration. The seasonal cycles differed depending on the kind of game hunted but generally were as follows (cf. Bishop 1974, Ray 1974, Gillespie 1975, Jarvenpa 1980, Yerbury 1986). Winter was spent in small family groups either hunting (before European trading), or hunting and trapping (after European trade began). In the subarctic forest barren-ground caribou were an important resource in the winter and in the southwestern boreal forest wood buffalo were important. Throughout the southern boreal forest moose and deer were the primary large game. Summer was spent in bigger settlements usually near large lakes where fishing and opportunistic moose or deer hunting supplied their food needs. Fall allowed berry collecting and duck, goose and bear hunting. Spring saw intense beaver trapping and hunting of the returning waterfowl. In recent years all of these subsistence activities have been supplemented by wage earning and government treaty payments.

The use of fire by boreal forest Indians has been carefully and systematically studied only by H.T. Lewis. Lewis (1977, 1982) has used ethnographic studies to show that the Cree Indians of northern Alberta regularly and systematically burned certain habitats to improve game and plant resources at least as far back as the mid-1800s. These fires differed in seasonality, frequency, size and behavior from lightning-caused fires. The upland boreal forest was *not* systematically burned because of the difficulty in managing the fire spread and its limited resource benefits. Generally the areas burned were along the margins of lakes, sloughs, bogs, meadows and swamps. These fires were set mostly in spring and sometimes in late fall. The fires were of small flame length and were intended to kill shrubs and trees and encourage grass and herbaceous growth of value to wildlife. By burning in spring or late fall the surrounding shaded woods would have been wet or still snow-covered, thus controlling the fire spread. Trails in swamps were also kept open by spring burning. These spring fires were not necessarily annual but as frequent as was required to keep the brush down. Thus, Indian fires had *specific* purposes and were set at *specific* times, were of low intensity and small in size.

Significantly, the Indians burned habitats which tended not to burn during the lightning fire season (June–August) and the fires were much smaller in size than lightning-caused fires. In short, they were management burns with specific objectives and a clear understanding of the effect desired. The lightning fire season was considered a foolish time to burn since fires could not be controlled (Lewis 1982).

Given the vegetation similarity of the boreal forest and the similarity of subsistence life ways between Indians in the boreal forest, Lewis's studies of the use of fires by the Cree of northern Alberta may give a general view of Indian use of fire in the boreal forest. However, it may be that regional and tribal differences did play a more important role in their use of fire. Further, the change starting in the 1500s from subsistence hunting to hunting and trapping, the population movements, and the population reduction caused by disease (in many cases reductions by as much as three-fourths: cf. Heidenrich and Ray 1976) all could have resulted in changes in the numbers and kinds of Indian-caused fires. At present, we do not know.

Lightning-caused fires

Lightning is the most significant cause of fires in the boreal forest accounting for about 90% of *the area* burned (Hardy and Franks 1963, Requa 1964, Barney 1969, 1971, Stocks 1974, Johnson and Rowe 1975, Stocks and Street 1983). Lightning-caused fires decline northward (e.g. Figure 2.2. for Ontario) generally as the frequency of thunderstorm days declines (Figure 2.3).

Three areas in the southern boreal forest are known to have higher lightning occurrence and more lightning-caused fires: the Sault Ste Marie–Sudbury, Red Lake–Kenora–Quetico Park areas of Ontario (Chapman and Thomas 1968) and north central Alberta (Harvey 1977). These areas appear to be associated with increased cyclogenesis. The lightning locating system (Krider *et al.* 1980) which now covers large areas of the boreal forest should offer greater possibilities for examining the relative importance of the density of lightning strikes and fuel moisture to the number of lightning fires.

Airstreams and the fire season

The seasonal variation in climate of the boreal forest is controlled by the interaction of the Arctic, cool Pacific, mild Pacific and North Atlantic airstreams (Hare and Hay 1974). The Rocky Mountains block the westerly flow of the Pacific airstream from penetrating uninterrupted over the eastern extent of the boreal forest. Only the boreal forest in the interior of Alaska is exposed to the Pacific airstream directly and then only during the summer.

For the boreal forest east of the Rockies, the cold, dry Arctic airstream

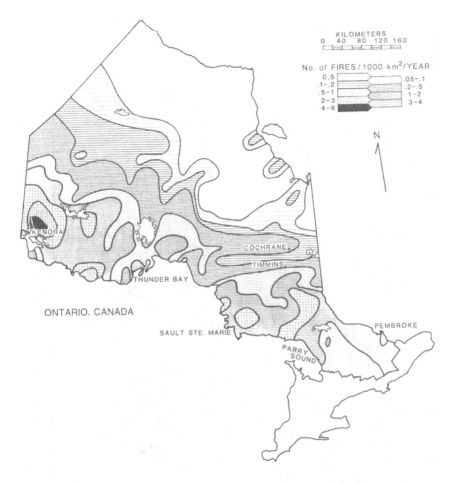

Figure 2.2. Pattern of lightning-caused fires in Ontario, Canada from April to September for 1965–76. (cf. Stocks and Hartley 1979).

flows south during the winter when radiation input into high latitudes is at its lowest and snow cover increases the albedo (Hare and Ritchie 1972). By February, this airstream usually covers all of the boreal forest (Figure 2.4). With the increasing radiation of spring and melting of the snow cover, the Arctic airstream retreats and is gradually replaced by Pacific and North Atlantic airstreams. Summer, the fire season in the boreal forest, finds the Arctic airstream positioned near the boreal forest–tundra boundary (Bryson 1966). A large wedge of mild Pacific air (Figure 2.4) occupies the boreal forest with varying frequency as far east as James Bay (Figure 2.5)

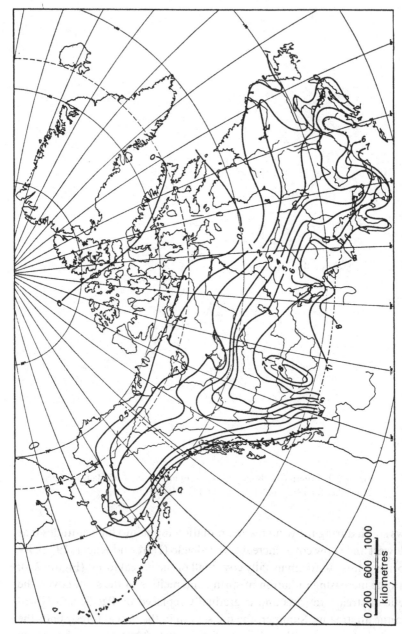

Figure 2.3. The mean number of days of thunderstorms in July for the period in Canada 1941–60 (cf. Kendall and Petrie 1962) and in Alaska 1931–60 (cf. Hare and Hay 1974).

0 200   600   1000

kilometres

with the eastern boreal forest in Ontario, Quebec and Labrador covered by a mixture of Pacific and North Atlantic airstreams (Figure 2.4).

The airstream patterns in the interior of Alaska are somewhat different. Here the boreal forest occupies the basin and plateaux of the Yukon River and its tributaries, the forest being wedged between the Brooks Range to the north, the Alaskan Range to the south and the MacKenzie Mountains (a part of the Rocky Mountains) to the east. During the winter the Arctic airstream dominates until the summer when the cool Pacific airstream replaces it. The cool Pacific air from the Bering and Chukchi Seas and the proximity of Arctic air north of the Brooks Range give the summer a continuous series of weak cyclones which results in high summer precipitation. The Pacific air also causes a slight increase in summer precipitation in the Yukon, MacKenzie and Keewatin Districts of the Northwest Territories although the intervening Ogilvy and MacKenzie Mountains create a minor rain shadow.

The fire season in the boreal forest is determined by the summer reorganization of atmospheric circulation over the boreal forest, switching from a colder stable Arctic airstream to warmer, unstable air caused by the deep intrusion of Pacific and Tropical airstreams. The beginning and end of the fire season in the boreal forest are related to the position of the Arctic airstream. Forest fires are possible once the Arctic airstream has retreated north. This is because the warmer, unstable Pacific or North Atlantic airstreams are more effective at drying fuels, owing to their much higher temperatures, and generate more lightning as an ignition source. Consequently, 12 minus the duration of the Arctic airstream in Figure 2.4 gives the number of months during which lightning fires can occur. Another indicator of when fires can occur is the date when mean air temperature is above 0 °C (Figure 2.6). The timing of this event is tied to the retreat of the Arctic air and increasing frequency of Pacific and Atlantic airstream incursions. Winter snow cover melts rapidly after the mean air temperature is above 0 °C (Hare and Hay 1974), allowing drying of surface fuels.

An example of this airstream-caused progression of fire starts can be seen in Figure 2.7 for an area east of Great Slave Lake. In the spring, ignitions move in a wave, northeast towards the treeline following the retreat of Arctic air. In July fires occur throughout the region but only rarely in the tundra (see Wein 1976). After August the Arctic airstream again begins to flow southward. Fires already burning north of the Arctic front may continue to burn but rarely do new fires start in areas under the influence of the Arctic airstream. A similar pattern has been reported in northern Quebec by Payette *et al.* (1989).

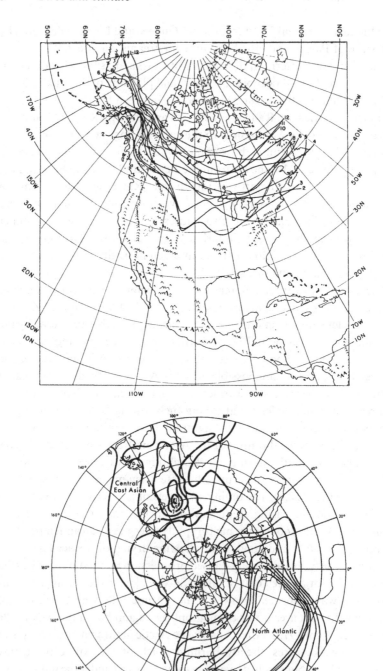

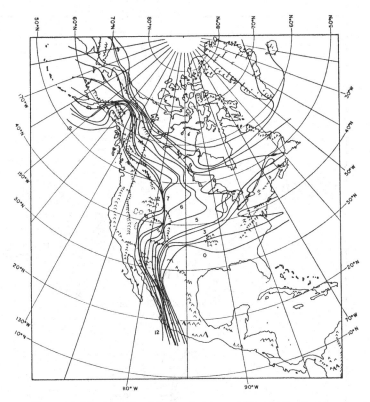

Figure 2.4. The duration of the Arctic (above, left), mild Pacific (above) and North Atlantic (below, left) airstreams over North America in months (cf. Hare and Hay 1974, Wendland and Bryson 1981).

Critical synoptic weather

Up to now we have been examining the time averaging of the circulation over several months that produces the strong seasonality of fires in the boreal forest, but not the synoptic disturbances of the circulation operating at the scale of several days to several weeks. It is at this scale that fires actually start, spread and are extinguished.

Large and widespread fires do not occur every year in the boreal forest but only when dry fuels, ignition and wind occur in sequence. During the 1980–9 forest fire seasons (Figure 2.8) such critical fire conditions developed often, allowing hundreds of large fires to burn in the boreal forest of Canada (Stocks 1983). Seventy percent of these large fires were caused by lightning. The conditions during these years were not extraordinary, the critical weather simply lasting longer and encompassing a larger area.

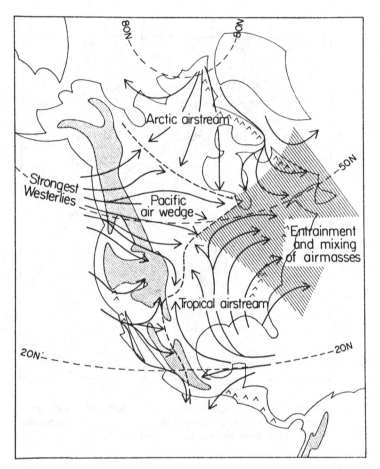

Figure 2.5. General characteristic of the flow of air streams over North America in July during the middle of the fire season in the boreal forest (cf. Bryson and Hare 1974).

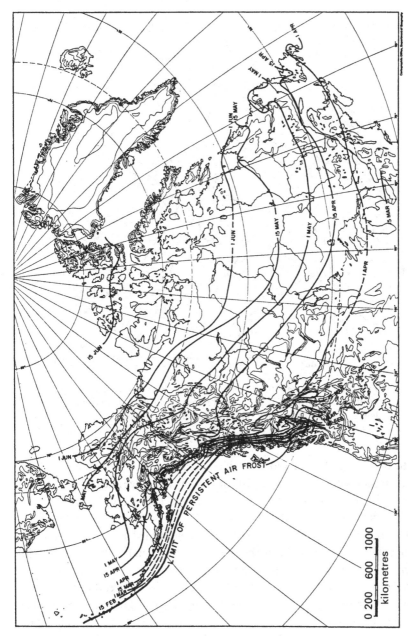

Figure 2.6. The mean date at which the mean air temperature reaches 0 °C. After this date snow is generally melting from the uplands although ice may still be on the lakes (cf. Hare and Hay 1974).

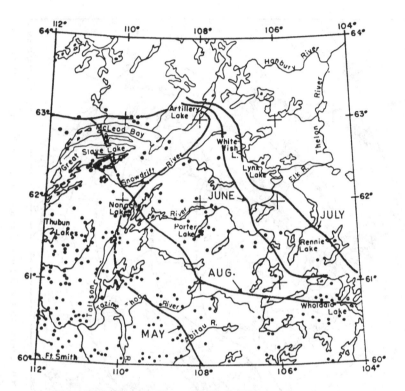

Figure 2.7. Lines mark the monthly limits of fires as they advance
northeastward from May to July and retreat southwestward in August. Tree-
line is approximately at the July line. Dots are the locations of fires from 1966
to 1971 (cf. Johnson and Rowe 1975).

Critical fire weather in most of the boreal forest is associated with a
characteristic persistent 50 kPa (500 mb) longwave ridge and its breakdown
(Schroeder 1950, Schroeder *et al.* 1964, Newark 1975, Stocks and Street
1983, Alexander *et al.* 1983).

The boreal forest synoptic pattern is determined by a succession of
tropospheric disturbances in the westerly flow usually along the Arctic
front. If the westerly flow is strong (zonal flow: see Figure 2.9), then these
traveling disturbances lead to alternations of warm–dry weather during the
upper level ridges in the flow and cool–wet weather during the upper level
troughs. However, at times this sequence is interrupted by ridges which stall
or are very slow moving (Figure 2.9). These ridges at the 50 kPa level are
anticyclones (highs) at the surface. They bring periods of warm–dry
weather which are caused by the subsiding, adiabatically warmed and dried

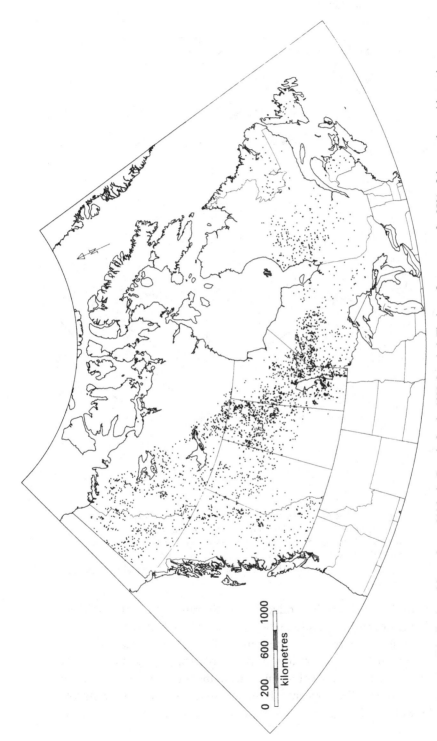

Figure 2.8. Distribution of forest fires > 200 ha in Canada for the 1980–9 period. The large fires account for 97% of the area burned but only 3% of the fires (source: B.J. Stocks, Forest Fire Research Unit, Forestry Canada, Ontario Region).

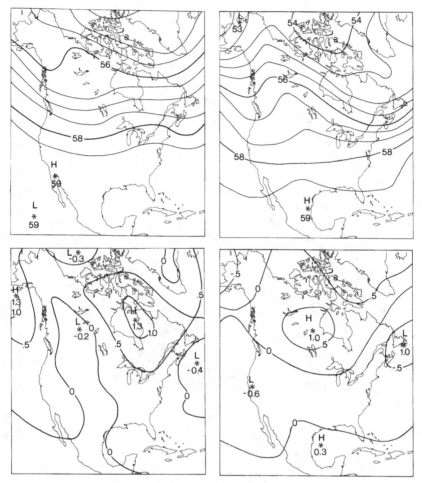

Figure 2.9. The mean 50 kPa (500 mb) height for June 1975 (above, left) and the deviation from normal (below, left), during a period of zonal flow over North America. The mean 50 kPa (500 mb) (above, right) and the deviation from normal (below, right) over North America for May 1980. An anomalous ridge is positioned over the central boreal forest resulting in meridional flow (cf. Stocks 1983).

air. In the boreal forest these ridges lead to the rapid drying of fuels because of high temperatures, low humidity, and usually light winds.

These ridges interrupt the alternation of surface cyclones and anticyclones and direct the westerly flow north or south of the blocking ridge. This results in meridional flow of warm air from the south into the boreal forest and cool air from the north into the grasslands and deciduous forests. Most of these blocking ridges are of short duration, lasting only a few days, but a

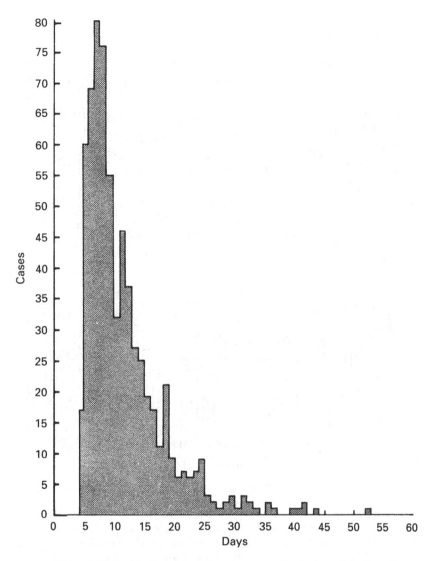

Figure 2.10. The Northern Hemisphere frequency distribution of blocking ridge durations for 1945–77 (cf. Treidl *et al.* 1981).

few last over a month (Figure 2.10). Harrington and Flannigan (1987; Flannigan and Harrington 1988) found that the area burned in the boreal forest of Canada increased with duration of the dry spells, with longer dry spells related to the duration of the blocking ridges. Blocking ridges tend to develop over Alaska, the Yukon and Northwest Territories during the fire season. The center of origination moves eastward from spring to summer.

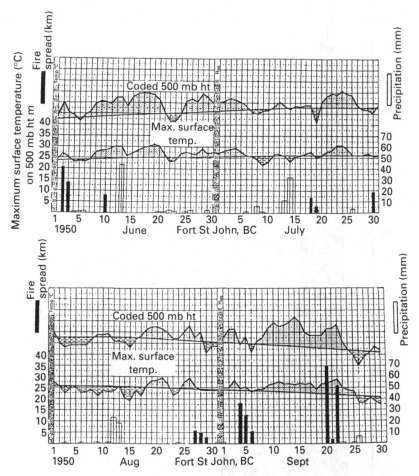

Figure 2.11. The coded 500 mb (50 kPa) pressure surface anomaly chart for Fort St John, British Columbia from June to September 1950 (cf. Murphy and Tymstra 1986). The 500 mb pressure height is plotted with the first and last digits of height removed. For example, 2 June has a height of 5500 m (MSL) and is plotted as 50. The chart gives the mean height of the 500 mb level and the positive (ridge) and negative (trough) anomalies. The surface temperature anomalies, precipitation and fire spread are also shown. The surface temperature generally mirrors the 500 mb pressure surface because of the proportionality of pressure and surface temperatures by the equation of hydrostatic equilibrium.

Most Northern Hemisphere blocking ridges develop in spring, with fewer during the summer (Treidl *et al.* 1981). If the ridges move it is usually in a southeasterly direction, carrying them across the center of the boreal forest.

The best way to understand the critical fire weather sequence is to use a 50 kPa anomaly chart (Figure 2.11). These charts identify the occurrence of ridges and troughs at the 50 kPa level (about 6000 m) above a particular region.

Blocking ridges weaken when troughs in the westerly flow are able to push the ridge eastward out of a region. Often these troughs are only a temporary disturbance and the ridge rebuilds (see Figure 2.11). These troughs are represented at the surface as cold cyclones which have lightning and precipitation. A complete breakdown of a ridge may reestablish the zonal flow with alternations of cyclones and anticyclones.

The critical fire weather sequence as seen in a 50 kPa chart has two stages (Nimchuk 1983). (1) The blocking ridge establishes itself, causing the fuels to dry rapidly. Fires during this phase will generally be man-caused since there is little or no lightning associated with the subsiding air flow of the ridge. Fire spread is slow during this stage because wind is light. Higher elevations can be drier than lower elevations due to the air mass subsidence. (2) The ridge may partially or completely break down into a trough. It is during this breakdown period that fires are started by lightning and/or have rapid spread. The high and variable winds in front of or behind the passage of the surface cold cyclone associated with the upper level trough are responsible for fires exhibiting their maximum spread rate. All critical conditions at this point are now in concert: dry fuel, lightning ignition source and high, variable winds. Brotak and Reifsnyder (1976) found that in 52 major wildland fires in North America, 54% followed dry cold front passage (no precipitation at the surface), and 20% preceded cold front passage. Figure 2.12 shows the location on the surface weather map where rapid fire spread can be expected.

The 1.4 million ha Chinchaga River Fire in northern British Columbia and Alberta (Murphy and Tymstra 1986) provides an excellent example of the periodicity of slow and rapid spread that is associated with the building and breaking down of upper ridges. Figure 2.13 gives the dates and areas of fire spread, with Figure 2.11 showing the correlation of rapid spread to ridge breakdown. Notice that the fire burns for the whole fire season, a usual occurrence in many fires in the boreal forest.

In summary, fire occurrence in the boreal forest has a general decreasing trend from its southern boundary towards its northern boundary. This is a

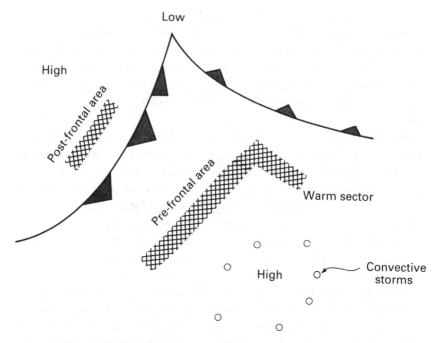

Figure 2.12. Idealized surface weather map showing the shaded areas where rapid and extreme fire behavior might be expected (cf. Brotak and Reifsnyder 1976).

result of the spring northward-retreat of the Arctic airstream that allows warmer unstable air to invade the boreal forest, causing increased lightning activity and generally warmer, drier conditions. The fall southward-readvance of the Arctic airstream ends the conditions suitable for fire ignition and spread. The alternation of tropospheric disturbances in the westerly flow over the North American boreal forest provides the sequences of fuel drying, lightning fire ignition and high winds that allow the forest to burn for long periods and produce large (relative to deciduous forest experience) fires. The relationship between time of fire ignition and length of time for fire spread needs to be researched. Are very large fires in the boreal forest mostly the result of early spring ignition and then a whole fire season of burning (e.g. the Chinchaga River fire: Figure 2.13)? Are decades such as the 1980s (Figure 2.8) associated with large-scale mid-tropospheric anomalies such as the Pacific North America Pattern or North Pacific and Atlantic Oscillations (Knox and Lawford 1990)?

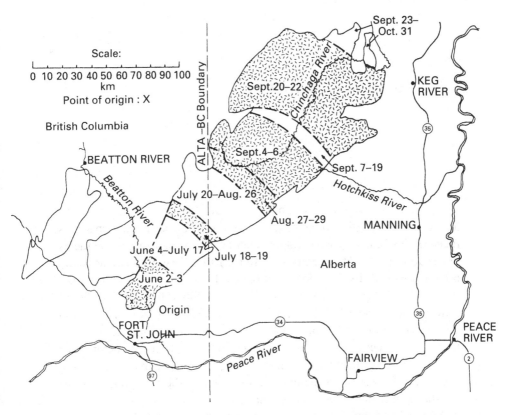

Figure 2.13. The reconstruction of the 1950 Chinchaga River fire. The fire burned for almost the whole fire season with a few periods in September of very rapid spread which occurred during the breakdown of upper ridges. Compare the periods of spread to the 500 mb (50 kPa) chart in Figure 2.11 (cf. Murphy and Tymstra 1986).

The next four chapters will consider in detail the mechanisms and variables in four fire behavior characteristics (rate of fire spread, fire intensity, duff consumption and fire frequency) and how they effect boreal population processes.

3

Forest fire behavior

A spreading forest fire is a complex combustion process in which the flaming front is heating and then igniting unburned woody and herbaceous fuels. In this heating process (Shafizadeh 1968), the moisture in the fuel is first evaporated (fuel temperatures > 100 °C), then the cellulose is thermally broken down and its breakdown products volatilized (> 200 °C) and finally the volatiles are ignited to form a visible flame (300–400 °C). The modes of heat transfer responsible for fire spread from the flaming front (Figure 3.1) are convection and radiation. Conduction does not contribute significantly to fire spread because wood and soil are such poor heat conductors. For conduction to be effective, the flame must be maintained in one place for a long time. However, to maintain a solid flame, the fire front must be constantly moving to recruit unburned fuels.

After the flaming combustion has ignited and burned most of the volatiles, the remaining carbon may burn as a solid by surface oxidation called glowing combustion. Flaming and glowing combustion are not discrete events in forest fires because of the complex mixture of fuel sizes, moistures and arrangements. However, the flaming front is dominated by combustion of gases and glowing combustion occurs primarily after it passes.

The differences between flaming and glowing combustion are of interest to ecologists because they can have different ecological effects. Flaming combustion is primarily responsible for plant death and glowing combustion for duff consumption and seedbed preparation. These effects will be discussed in detail in Chapters 4 and 5.

In this chapter, two issues will be discussed: the general mechanism and relevant variables of fire rate of spread, and why high rates of spread are so common in the boreal forest.

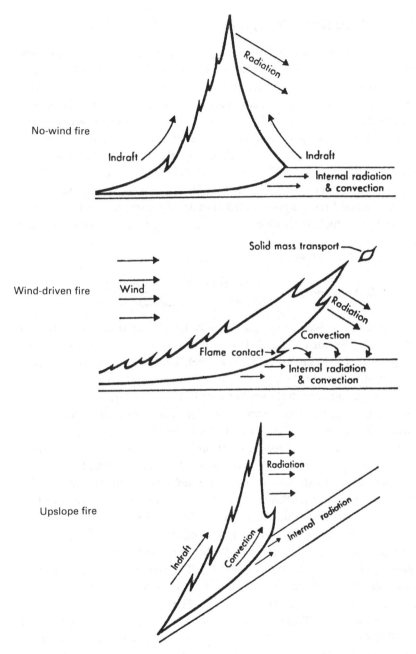

No-wind fire

Wind-driven fire

Upslope fire

Figure 3.1. Idealized flaming front with modes of heat transfer (cf. Rothermel 1972).

Mechanism of fire spread

A first principle understanding of wildfire rates of spread does not exist at present (Albini 1984), and if it did, ecologists would probably not find it useful with our meager understanding of fire effects. However, enough empirical and theoretical (mathematical and dimensional analysis) research has been done on forest fuels to give an understanding of the important variables and their general relationship in the rate of spread process in ground fires (e.g. Fons 1946, Hottel *et al.* 1965, Rothermel and Anderson 1966, Berlad 1970, Rothermel 1972, Pagni and Peterson 1973, Albini 1976) and crown fires (Albini 1981, 1985, 1986, Albini and Stocks 1986). Ground fires are propagated in fuels which rest on the forest floor. Crown fires, which are discussed in Chapter 4, spread in both the surface fuels and tree crowns.

A simple thermal model of fire spread is:

$$\text{Rate of spread} = \frac{\text{heat flux from the flaming front}}{\text{heat required for fuel ignition}} \qquad (3.1)$$

This equation suggests in a general way which variables play a role in determining the rate of spread. The 'heat flux from the flaming front' is the energy available to heat the unburned fuel. This will not be the total heat flux from the flame since some will be lost to the air, etc. This propagating flux is a function of both the heat given off by the fuel when it is combusted (joules per kilogram) and the rate at which a specific density of fuel is being burned (kilograms per area–time). Traditionally, slope and wind are entered as effects on the propagating heat flux since both cause the flame to have higher heat transfer by lessening the distance from the flame to the unburned fuel (Sheshukov 1970, Rothermel 1972, Van Wagner 1977b). Slope and wind are coefficients which increase the heat flux relative to a level surface with no wind (Van Wagner 1977b, 1988b).

The 'heat required for ignition' is a function of the thermal physical and chemical properties of the fuel and the heat required to raise this fuel to its ignition temperature (usually $> 300\,^\circ\mathrm{C}$). The physical properties include measures of the bulk density, mass, surface/volume ratio and the arrangement of the fuel on the ground. The chemical properties consist most significantly of the moisture content. All other chemical properties are usually relatively constant compared to the large and variable heat sink of water (see Chapter 4).

I have specifically not written mathematical equations for the compo-

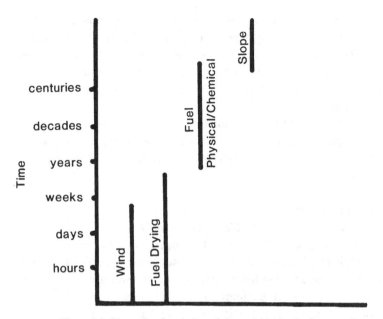

Figure 3.2. The scale of variation of the variables in the fire rate of spread process. The physical and chemical variables do not include water which is in fuel drying.

nents of rate of spread since this would divert us into technical details unnecessary for our argument here. The reader should turn to any of the papers mentioned at the beginning of this section for this detail. The Rothermel (1972) paper in particular is readable and has been widely used by the US Forest Service in its fire behavior models (Albini 1976).

Fuel types

Expansion of the rate of spread equation (equation 3.1) requires considerable knowledge of the physical and chemical characteristics of the fuel, wind and slope. The strategy for simplifying the rate of spread equation involves understanding which factors vary the most and which vary the least. The relatively constant factors can then be collected into a few stylized types and the highly variable factors set for specific conditions. Figure 3.2 shows that while wind and fuel drying vary over a short time span, other fuel characteristics and slope are relatively stable. Slope is easily and quickly estimated and hence is usually put with wind and fuel wetting and drying variables. This leaves non-drying fuel characteristics which are reasonably stable but very time-consuming to measure. Consequently, fuel types arise from the insensitivity of their variables in the governing equations. Fuel types are simplifications of the rate of spread process because slight differences between forest stands in fuel characteristics are insignificant when compared with the large short-term variations in the weather (wetting and drying) variables.

The Canadian Forest Fire Behavior Prediction System (Alexander *et al.* 1984, Van Wagner 1973b) recognizes seven *natural* fuel types in the boreal forest. These types are described below and in Table 3.1 directly from the Fire Behavior System.

Coniferous group
Fuel Type C-1 (Spruce–Lichen Woodland)

This fuel type is characterized by open, park-like black spruce stands occupying well-drained uplands in the subarctic. Jack pine and white birch are minor associates in the overstory. Forest cover occurs as widely spaced individuals and dense clumps. Tree heights vary considerably but bole branches (live and dead) uniformly extend to the forest floor and layering development is extensive. Woody fuel accumulation on the forest floor is very light and scattered. Shrub cover is exceedingly sparse. The ground surface is fully exposed to the sun and covered by a nearly continuous mat of ground-dwelling lichens averaging 2–4 cm in depth above mineral soil.

Fuel Type C-2 (Boreal Spruce)

This fuel type is characterized by pure, moderately well-stocked black spruce stands on lowland (excluding *Sphagnum* bogs) and upland sites. Tree crowns extend to or near the ground and dead branches are typically draped with lichens. The flaky nature of the bark on the lower portion of stems boles is pronounced. Low to moderate volumes of downed woody material are present. *Ledum groenlandicum* is often the major shrub component. The forest floor is dominated by a carpet of feather mosses (*Pleurozium schreberi, Hylocomium splendens, Ptilium crista-castrensis,* and *Dicranum* spp.) and/or ground-dwelling lichens (chiefly *Cladonia*). *Sphagnum* mosses may occasionally be present but they are of little hindrance to surface fire spread. A compact organic layer often exceeds a depth of 20–30 cm.

Fuel Type C-3 *(Mature Jack Pine)*

This fuel type is characterized by pure, fully stocked (1000–2000 stems ha^{-1}) jack pine stands that have matured at least to the stage that crown closure is complete and the base of live crown is substantially separated from the ground. Dead surface fuels are light and scattered. Ground cover is basically feather moss over a moderately deep (*c.* 10 cm) compact organic layer. A sparse conifer understory may be present.

Fuel Type C-4 *(Immature Jack Pine)*

This fuel type is characterized by pure, dense jack pine stands (10 000–30 000 stems ha^{-1}) in which natural thinning mortality results in a large quantity of standing dead stems and dead and down woody fuel. Vertical and horizontal fuel continuity is characteristic of this fuel type. Surface fuel loadings are greater than in Fuel Type C-3; organic layers are shallower and less compact. Ground cover is mainly needle litter, partially elevated and suspended within a low (*Vaccinium* spp.) shrub layer.

Fuel Type C-5 *(Red and White Pine)*

This fuel type is characterized by mature stands of red and white pine in various proportions, sometimes with small components of white spruce, and white birch or balsam fir. A shrub layer, usually *Corylus cornutas*, may be present in moderate proportions. The ground surface cover is a combination of herbs and pine litter. The organic layer is usually 5–10 cm deep.

Deciduous group
Fuel Type D-1 *(Leafless Aspen)*

This fuel type is characterized by pure, semi-mature trembling aspen stands prior to 'green-up' in the spring or following leaf fall and dieback of lesser vegetation in the autumn. A conifer understory is noticeably absent but a well-developed medium to tall shrub layer is typically present. Dead and fallen roundwood fuels are a minor component of the fuel complex. The principal fire-carrying fuels consist chiefly of deciduous leaf litter and cured herbaceous material which are directly exposed to wind and solar radiation. In the spring, the duff mantle (F and H horizons) is seldom available for combustion because of its generally high moisture content.

Table 3.1. *Brief summary of Canadian Forest Fire Behavior Prediction System Fuel Type characteristics* (Alexander *et al.* 1984)

Fuel type	Forest floor and organic layer	Surface and ladder[a] fuels	Stand structure and composition
C-1	Continuous ground dwelling lichen; organic layer absent or shallow, uncompacted.	Very sparse/herb shrub cover and downed woody fuels; tree crowns extend to ground.	Open black spruce stands with dense clumps; assoc. spp. jack pine, white birch; well-drained upland sites.
C-2	Continuous feather moss and/or cladonia; deep, compact organic layer.	Continuous shrub, e.g. *Ledum groenlandicum* common; low to moderate downed woody fuels; tree crowns extend nearly to ground; arboreal lichens, flaky bark.	Moderately well-stocked black spruce stands on both upland and lowland sites; sphagnum bogs excluded.
C-3	Continuous feather moss; moderately deep, compact organic layer.	Sparse conifer understory may be present; sparse downed woody fuels; tree crowns separated from ground.	Fully-stocked jack or lodgepole pine stands; mature.
C-4	Continuous needle litter; shallow, moderately compacted organic layer.	Moderate shrub/herb cover; vertical crown fuel continuity; heavy standing dead; downed dead woody fuel.	Dense jack or lodgepole pine stands; immature.

C-5	Continuous needle litter; moderately shallow organic layer.	Moderate herb and shrub, e.g. hazel; moderate dense understory, e.g. red maple (*Acer rubrum*), balsam fir; tree crowns separated from ground.	Moderately well-stocked red and white pine stands; mature; assoc. spp. white spruce, white birch, trembling aspen.
D-1	Continuous leaf litter; shallow uncompacted organic layer.	Moderate medium to tall shrubs and herb layers; absent conifer understory; sparse dead and downed woody fuels.	Moderately well-stocked trembling aspen stands; semi-mature; leafless (i.e. spring and fall).
M-1 and M-2	Continuous leaf litter in deciduous portions of stands; discontinuous feather moss and needle litter in conifer portions of stands; organic layers shallow, uncompacted to moderately compacted.	Moderate shrub and continous herb layers; low to moderate dead and downed woody fuels; conifer crowns extend nearly to ground; scattered to moderate conifer understory.	Moderately well-stocked mixed stands of boreal conifers (black/white spruce, balsam fir) and deciduous species (trembling aspen, white birch). Fuel types are differentiated by season and percentage conifer:deciduous spp. composition.

Note:
[a]Ladder fuels are fuels which provide vertical continuity between the surface fuels and crown fuels in a forest.

Mixedwood group
Fuel Types M-1 (Boreal Mixedwood – leafless) and M-2 (Boreal Mixedwood – summer)

These fuel types are characterized by mixtures of coniferous and deciduous tree species in varying proportions: black spruce, white spruce, balsam fir, trembling aspen, and white birch. Within any combination, individual species can be present or absent from the mixture, i.e. one or both of the broadleaf species can occur, and one, two or all of the conifers. In addition to the diversity in species composition, stand mixtures exhibit wide variability in stand structure and development but are generally confined to moderately well-drained upland sites. Two phases associated with the seasonal variation in the flammability of the boreal mixedwood forest are recognized: leafless stage – spring and fall periods (Fuel Type M-1), and summer period (Fuel Type M-2). The rate of spread in both fuel types is weighted according to the proportion (expressed as a percentage) of softwood (S) and hardwood (H) components. In the summer, after the deciduous overstory and understory vegetation have leafed out, fire spread is greatly reduced with maximum spread rates possibly only one-fifth as fast as spring or fall fires under similar burning conditions.

Determining rate of spread

Since a fire spreads initially in ground fuels, any determination of spread rate requires consideration of both the ground fuel layer and the rapidly changing weather variables through a series of models of the wetting and drying processes in the ground fuel layers. The boreal forest floor can be divided into three layers which have three moisture drying rates called Codes in the Canadian Fire Weather Index (Van Wagner 1987). The actual equations of the Fine Fuel, Duff and Drought Moisture Codes and a comparison with the American National Fire-Danger Rating System (Deeming *et al.* 1974) can be found in Van Wagner (1987).

The Fine Fuel Moisture Code gives the (Coded) moisture content of litter, twigs, needles and other dead fuels less than 2 cm in diameter directly on the forest floor. These fuels dry out very quickly, taking about two-thirds of a day to lose two-thirds of their free moisture at equilibrium when the air is at 20 °C and 40% relative humidity ('equilibrium moisture'). In the Fine Fuel Code, the free moisture at equilibrium varies with temperature and particularly relative humidity. This layer has a very low water holding capacity which, along with its high surface/volume ratio, explains its rapid drying. This Code assumes, as do all other Codes discussed, that drying is an exponential process (see Chapter 4). Wetting does not initially affect the fuel moisture up to 0.5 mm of precipitation, then has a progressively large effect up to 5.7 mm of precipitation, and then less and less. The required weather parameters are temperature, humidity, wind speed and rainfall in the last 24 hours. All parameters are measured at or near solar noon. This time was chosen because fires are generally most active in the afternoon.

The Duff Moisture Code estimates the moisture of the F layer of the soil humus layer. This layer, because of its more compact geometry, takes about 12 days to dry to two-thirds of its equilibrium moisture. The weather parameters used to predict moisture are temperature, humidity and rain. Wind is no longer needed since its effect on drying is small.

The Drought Code estimates the moisture of the deepest and most compact duff layer and large downed logs. This fuel layer takes 52 days to dry to two-thirds its equilibrium moisture.

The predicted fire spread is determined by combining the Fine Fuel Moisture Code, which gives the availability of this fast drying fuel, and the wind or slope which increases the spread by tipping the flame closer to the fuel. The predicted spread is called the Initial Spread Index since it does not include the heavier fuels which may be involved in larger or established fires. Figure 3.3 gives the Initial Spread Index as a function of Fine Fuel Moisture Code and wind. Notice the rapid increase in initial spread with increasing wind speed. Higher codes indicate lower fuel moisture.

Finally, the Initial Spread Index was correlated to actual spread in specific fuel types in over 200 experimental burns and wild fires. Figure 3.4 shows the experimental and wildfire data used in constructing the rate of spread for the Mature Jack Pine Fuel Type.

Why are there high rates of spread in the boreal forest?

Figure 3.5 gives the empirically determined rates of spread and meteorologically determined Initial Spread Index for the major boreal fuel types. Note that as the deciduous components in the fuel types increase, the rate of spread for a specific set of weather variables in the Initial Spread Index decreases. The exception to this is red–white pine (C-5) which, because of its high crowns (> 20 m) and sparse ground fuels, has a rate of spread as low as aspen (D-1).

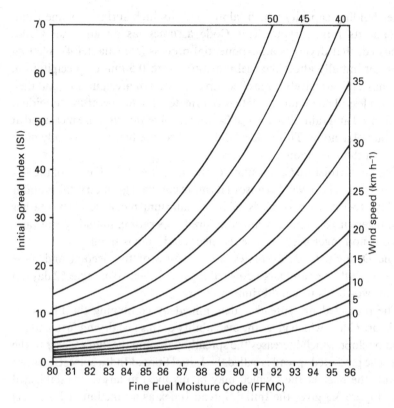

Figure 3.3. The Initial Spread Index as a function of the Fine Fuel Moisture Code and 10 m open, level terrain wind speed in the Canadian Fire Weather Index (from Van Wagner 1987).

The lower graph on Figure 3.5 shows the empirical values at which, on average, active crown fires can be expected. Active crown fires are where flames are propagated from the ground fuels up into the tree crowns (for more details, see Chapter 4). Again, the deciduous fuels require higher rates of spread for crowning to occur. Immature jack pine (C-4) has the lowest rate of spread at crowning and mixedwood (M-2) the highest. Boreal spruce (C-2) has lower crowning than mature jack pine (C-3), primarily because of the greater ground fuels and presence of dead branches nearer the ground which provide easier access of the flames into the canopy. Thus, as a first approximation, the preponderance of coniferous fuels is significant in explaining the high rates of spread observed in the boreal forest.

What is it about the coniferous fuel types that contributes to their high

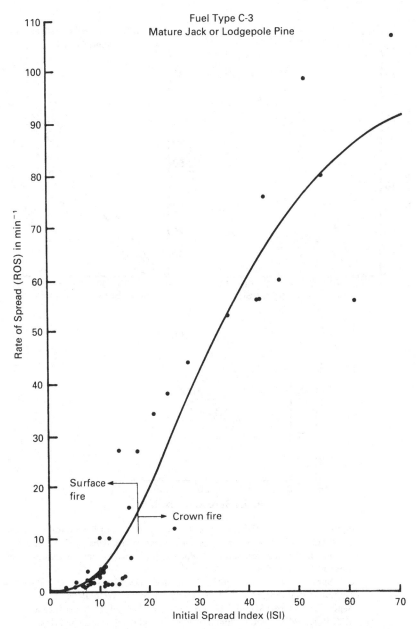

Figure 3.4. The empirical measurements of rate of fire spread in experimental fires and wild fires (dots) compared to the predicted Initial Spread Index values (line) based mature jack pine Fuel Type C-3 and weather variables at the time of the fire (from Lawson *et al.* 1985). Notice the values above which crown fires occurred.

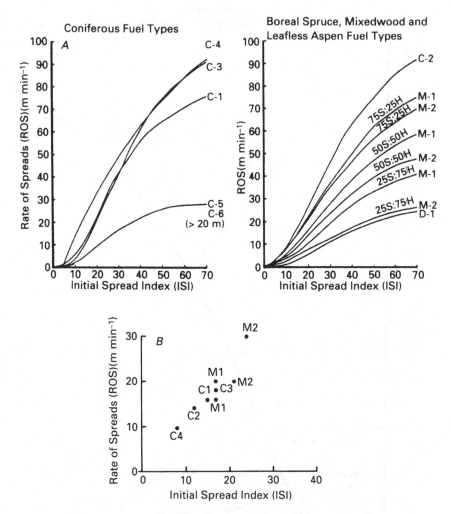

Figure 3.5. *A.* The Initial Spread Index versus the Rate of Spread for coniferous and deciduous Fuel Types in the Canadian Forest Fire Behavior Prediction System. *B.* The empirical values at which active crown fires can be expected. (cf. Alexander *et al.* 1984).

C-1, Spruce–Lichen Woodland; C-2, Boreal Spruce; C-3, Mature Jack Pine; C-4, Immature Jack Pine; C-5 and C-6 (> 20 m), Red and White Pine; D-1, Leafless Aspen; M-1, Boreal Mixedwood – leafless: % softwood (S), % hardwood (H); M-2, Boreal Mixedwood – summer: % softwood (S), % hardwood (H).

rates of spread? Although the low decomposition rates of boreal conifer stands lead to accumulation of a significant forest floor of dead biomass when compared with deciduous stands (Bray and Gorham 1964, Vogt *et al.* 1986), this in itself cannot be completely responsible for the high rate of spread. Conifers do, however, produce a greater abundance of fine fuels in the form of needles, small twigs, resinous products and small bark flakes than do most deciduous trees. Also, lichens and mosses are common on the forest floor. These live fuels dry as rapidly as dead fuels (Mutch and Gastineau 1970). Further, conifers retain more dead branches of all sizes than do deciduous trees. The crown shape of conifers allows easy access of flame from the ground into the crown. When this conical shape is not present (e.g. in red and white pine), crowning and high rates of spread require drier and windier conditions. Finally, the moisture content of deciduous tree foliage is > 150% dry weight while that of conifers is usually 100% or lower (Van Wagner 1967).

One of the significant empirical relationships which has emerged in fire behavior research is that forest fuels can be arranged into a few associations of fuel elements. These 'fuel types' have similar fire behavior, particularly rates of spread (Figure 3.5). In the boreal forest, it is the combination of weather (fuel drying) and conifer fuel types which accounts for the high rate of fire spread.

Fire size and shape on tree seed dispersal distances

The distinctive 'elliptical' shapes of high rate of spread fires (Figure 3.6) compared to the more 'circular' shape of low rate of spread fires (Van Wagner 1969b, Anderson 1983, Alexander 1985) may have effects on seed dispersal distances of trees. All the tree species of the North American boreal forest (both gymnosperms and angiosperms) possess wind dispersed seeds. This is atypical of forests in general, e.g. usually less than 10% of tropical tree species are wind dispersed (Van der Pijl, 1972). Figure 3.7 shows observed wind dispersal curves for various spruce species dispersing into large clearings (burns or clear-cuts). The steep decline in seed density in the lee of the forest edge is a result of a lower wind speed adjacent to the forest (a shelter effect: Nageli 1953, Greene 1989). A micrometerological model of dispersal of seed by wind from an area source shows that seed-fall density at 200 m distance from a forest edge should be only about 1–4% of the seed-fall density in the forest (Greene 1989). Consequently, as can be seen in Figure 3.7, dispersal from the edge into a large burn (e.g. at

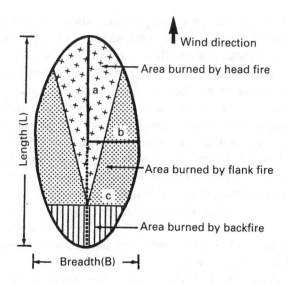

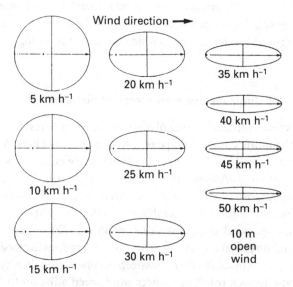

Figure 3.6. Simple elliptical fire growth model proposed by Van Wagner (1969a) and effects of increasing wind velocity on shape and rate of spread.

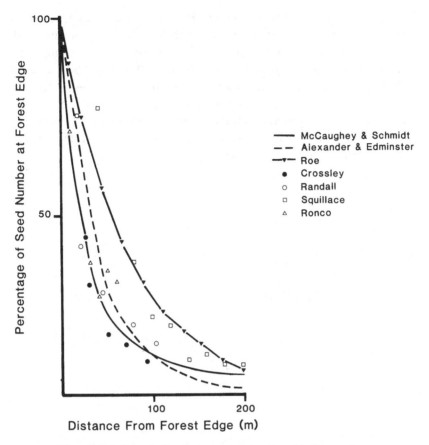

Figure 3.7. Relative decline in spruce seed densities with distance for seven studies (cf. Greene 1989).

diameters greater than 400 m) will be limited. It has long been understood by foresters that inadequate regeneration densities of large clear-cuts is a direct result of limited dispersal (Siggins 1933, Franklin and Smith 1973, 1974). Serotinous and semi-serotinous coned species such as jack pine and black spruce are clearly at an advantage by having short dispersal distances even in large burns. This, of course, presumes they were present and reproductive in the pre-fire forest. Non-serotinous species such as white spruce, balsam fir, ground juniper, red pine, birches and trembling aspen will have colonization limitations unless they have surviving reproductive individuals scattered throughout the burn. Several studies have shown juniper (Diotte and Bergeron 1989) and red pine (Van Wagner 1971b,

Bergeron and Brisson 1990) to be restricted by dispersal distances after local destruction by fire. At present, there are few studies which have tried to collect empirical data of the density, size and dispersion of surviving trees or tree patches in boreal forest fires. Eberhart and Woodard (1987) studied 69 burns ranging in size from 21 to 17 770 ha. The smallest fires tended not to have unburned islands while the number of unburned islands per hectare was highest in the largest fires. If this pattern is common in the boreal forest, it is possible that non-serotinous tree species may not have the serious colonizing problem they at first seem to have.

4

Fire intensity

Fires in the boreal forest are characterized by high intensities which result in extensive mortality in canopy and understory plants. This chapter will examine the heat transfer from a flaming fire front (fire intensity), how the high intensities in the boreal forest come about and how these intensities can be coupled to the effects on plants.

The flame is the essence of a fire to most people. To characterize this flame, the rate at which heat is given off by the flame (**fire intensity**) is more useful than the flame's temperature. Temperature is a quantification of the degree of hotness of a body, while heat is the quantification of the work transferred from a body at higher temperature to one at lower temperature. The temperature of a single burning twig can be the same as a large crown fire yet clearly the crown fire is transferring more heat from the flaming front to the immediate environment than is the burning twig. It is this heat transfer (intensity) which causes the adjacent fuels to be heated and burn, thereby releasing more heat and propogating the fire. Also, plant death and injury will be dependent on the heat transferred to them and on how much of this heat is absorbed so as to raise the plant's temperature to the lethal level.

The concept

Combustion of forest fuels involves pyrolysis, the chemical decomposition of the fuel by the action of heat, followed by ignition. Two types of combustion are recognized, flaming and glowing (Shafizadeh and De Groot 1976). Flaming combustion involves the fuel's extractive and holocellulose chemical components. Extractives (terpenes, fats, oils and waxes) are low molecular weight compounds which have high heats of combustion (heat released per unit mass) and are volatilized at low temperatures. Cellulose, on the other hand, decomposes at somewhat higher temperatures to produce volatile products which then burn with a visible flame. Ligin components are very stable when heated and produce a carbonaceous compound called char. Glowing combustion is the surface oxidation of char.

Forest fires spread primarily by heat transfer from flaming combustion. Consequently, fire intensity attempts to characterize the heat flux from the flame by the equation

$$I = c \frac{dm}{dt}$$

where c is the heat released per unit mass by flaming combustion (heat of combustion), dm/dt is the rate of mass loss of the fuel. For forest fires the mass of the fuel consumed per unit area is substituted for dm and rate of spread for $1/dt$.

Consequently, fire intensity (I) is usually given as

$$I = cmr \tag{4.1}$$

where m is the mass of the fuel consumed by the flaming front and r is the rate of spread (cf. Byram 1959).

The dimensions of each of these variables, including the rate of heat released, I, show how these variables are put together in the equation of fire intensity. The rate of heat released by a flaming front is by definition the energy per length of fire front per unit time. Energy has the dimensions ML^2T^{-2} (M is mass, L is length and T is time) which in SI units is joules. The rate of heat release is thus joules per time · length (MLT^{-3}). The weight of fuel consumed is mass/area or ML^{-2} and in SI units kg m^{-2}. The heat released per unit mass of fuel (heat of combustion) is energy/mass or $(ML^2T^{-2})/M$ and in SI units joules kg^{-1}. Finally, the rate of spread of the fire into new fuel is length per time (LT^{-1}). The SI units are:

$$\text{kW m}^{-1} = \text{kJ kg}^{-1} \cdot \text{kg m}^{-2} \cdot \text{m s}^{-1}$$

A kilowatt is equivalent to 1 kilojoule per second.

Some confusion has arisen about the dimensions of fire intensity (Tangren 1976, Van Wagner 1977c). Fire intensity will be used here to mean intensity *per length* of fire front. Perimeter fire intensity has also been used, but this averages the intensity from the higher fire intensity at the head of the fire and the lower fire intensities of the flank and rear. Total fire intensity is the maximum rate of heat output of the whole fire and is of little ecological use. Perimeter and total fire intensities will not be used here, but readers should watch the dimensions of intensities in the literature.

The heat transmission which goes into fire intensity can take three modes: conduction, radiation and convection. Consequently, fire intensity could be calculated for each of these modes separately or all together. In actual forest fires it is presently impossible to separate these modes of heat transfer. Nelson and Adkins (1986) discuss the empirical problems in calculations and measurements of intensity by conductive, convective and radiative transfer.

We will now discuss each of the variables in the fire intensity equation 4.1 and what values they can be expected to take in the boreal forest. The variation in these values will help us determine which variables are more important in determining the magnitude of fire intensity.

Heat of combustion (c)

The *amount* of heat released per unit mass is called the heat of combustion (c). In the past the heat of combustion was measured in calories per gram of fuel but is now given in SI units as joules per kilogram [$4.1836 \text{ cal g}^{-1} = 1 \text{ kJ kg}^{-1}$]. The *high* heat of combustion is the maximum heat that could be released by a *dry* fuel if it were combusted completely (both flaming and glowing combustion) to water and CO_2. The *low* heat of combustion (sometimes called heat yield) adjusts the high heat of combustion down due to the fuel left after the flaming front has passed and before glowing combustion has started. The low heat of combustion is usually associated with the volatiles given off when the fuel is heated. Susott (1982) recognized three 100 °C wide temperature regions which characterized the volatiles evolved from forest foliage, wood, small stems and bark. Temperatures below 300 °C were dominated by extractives (ether and benzine/ethanol), 300–400 °C were dominated by polymers of hemicellulose, cellulose and lignin while above 400 °C the gas consisted of lignin and lignin-like components composed of aromatic polymers.

To predict the heat of combustion available to the flaming fire front, a simple heat budget (Susott *et al.* 1975) is solved for the heat of combustion of the volatiles in the early part of combustion (up to 500 °C):

$$
\begin{vmatrix} \text{heat of combustion} \\ \text{of the volatiles} \\ \text{up to } 500\,°\text{C} \end{vmatrix} = \begin{vmatrix} \text{heat of combustion} \\ \text{of the total fuel} \\ \text{up to } 500\,°\text{C} \end{vmatrix} - \begin{vmatrix} \text{heat of combustion} \\ \text{of the char} \end{vmatrix} \cdot \begin{vmatrix} \text{fraction} \\ \text{char} \end{vmatrix}
$$

$$
(\text{kJ kg}^{-1}) \qquad\qquad (\text{kJ kg}^{-1}) \qquad\qquad (\text{kJ kg}^{-1}) \qquad (\%)
$$

Table 4.1. *Mean values of high heat of combustion for some boreal forest species*

Species	Plant part	kJ kg^{-1}	Source
Empetrum nigrum	leaves	23 471	1,2
	stems	23 470	1
Ledum palustre	leaves	23 220	1
	stem	22 550	1
Abies balsamea	needles	22 986	3
Betula glandulosa	leaves	21 086	1
	stem	22 592	1
Pinus resinosa	needles, litter (summer)	21 824	3
Picea mariana	needles	21 816	4
Pinus banksiana	needles, litter (summer)	21 694	3
Ledum groenlandicum		21 668	2
Vaccinium vitis-idaea	leaves	21 588	1,2
	stem	21 421	1
Vaccinium uliginosum	leaves	20 751	1
	stem	21 295	1
Alnus crispa	leaves	20 877	1
	stem	21 379	1
Picea glauca	needles	20 556	4
Populus tremuloides	litter (fall)	20 822	1
Salix glauca	leaves	19 622	1
	stem	20 961	1
Stereocaulon paschale		19 312	2
Eriophorum vaginatum	leaves	18 618	1
	dead leaves	19 110	1
Sphagnum		18 618	1
Calamagrostis canadensis	leaves	18 576	1
	dead leaves	18 451	1
Epilobium angustifolium	leaves	18 492	1
	stem	18 241	1
Duff (*Pinus strobus*)		17 722	3
Cladonia alpestris		17 584	2

Sources: 1. Sylvester and Wein 1981, 2. Bliss 1962, 3. Van Wagner 1972b, 4. Chrosciewicz 1986.

where char is the fuel remaining after flaming combustion and available to glowing combustion.

Most wildland fuels in the boreal forest (Table 4.1) and other vegetation types (Susott 1982) have similar total heats of combustion and, consequently, an average value of 18 700 kJ kg^{-1} (Van Wagner 1972b, Albini 1976) or 21 400 kJ kg^{-1} (Susott 1982) is usually used no matter what the fuel. This is justified since the variation in heat of combustion between fuels is always within 10% of the mean (Susott 1982).

Fuels do, however, have some differences in the amounts of volatiles and char. This is particularly noticeable between dead fuels which, because of decomposition, usually have fewer volatiles than do live fuels. However, since the flaming front burns in an assortment of fuels, differences in volatile heat of combustion are usually averaged out (Van Wagner 1972b, Christensen 1985). Further, the differences between the mean volatile and char heat available to flaming combustion (i.e. temperature below 500 °C) and char fraction are small. Susott's data (1982) for 43 typical forest fuels from throughout North America give a mean volatile heat of combustion available to flaming of 12 700 ($\pm$1800) kJ kg^{-1}, mean char heat of combustion of 8700 ($\pm$2210) kJ kg^{-1} and mean char fraction for foliage of 28.1 ($\pm$2.37)%, wood 20.52 ($\pm$2.32)%, stems 25.88 ($\pm$2.48)% and bark 37.33 ($\pm$7.76)%.

Susott's data indicate that on average 60% of a fuel's heat of combustion is available to flaming and 40% to glowing combustion. Therefore, one can choose for the heat of combustion in the intensity equation, either the average total heat of combustion of 18 700 kJ kg^{-1} (e.g. Table 4.1) or the lower volatile heat of combustion available to flaming of 12 700 kJ kg^{-1} (e.g. Susott 1982). Although these differences in high and low heat of combustion may seem large for boreal fuels, the rank of fuel parameters affecting fire intensity are moisture, surface/volume ratio, bulk density, silica-free ash content and, lastly, the heat of combustion (Sylvester and Wein 1981). The higher extractive content may play a role in fire starts only where initiation of combustion at temperatures less than 300 °C would be important (Shafizadeh and DeGroot 1976).

Rate of spread (*r*)

The rate of spread (*r*) is the distance perpendicular to the moving flame front per unit time. It is not the increase in perimeter or length of the fire (Van Wagner 1965, 1977c, Tangren 1976). Strictly, the fire must have established a steady-state flow for the rate of spread to have meaning.

The mechanisms for the high rates of spread in the boreal forest have been discussed in Chapter 3. Rates of spread up to 100 m min^{-1} have been reported (Kiil and Grigel 1969) but values of 5–10 m min^{-1} are usual (Hardy and Franks 1963, Sando and Haines 1972, Stocks and Walker 1973, Stocks 1974, 1987b, Kiil 1975, Alexander *et al.* 1983, Lanoville and Schmidt 1985, DeGroot and Alexander 1986).

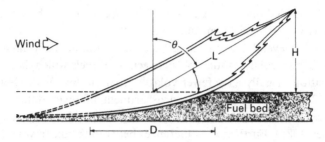

Figure 4.1. Convention for measuring surface flame length (L), flame height (H), flame tilt angle (θ), flame depth (D).

Fuel consumed (m)

The weight of the fuel consumed during the passage of the flaming front cannot generally be separated from that lost by glowing combustion. In the boreal forest, ground fires consume approximately 2–3 kg m^{-2} of fuel while crown fires consume 3–5 kg m^{-2} (Van Wagner 1965, Lawson 1973, Quintilio *et al.* 1977, Stocks 1987a).

Flame length and fire intensity

Of the three variables in fire intensity (equation 4.1), only rate of spread is easily measured in wild fires. This could create an almost insurmountable problem in using fire intensity, were it not that intensity is allometrically related to flame length. Average flame length is the distance between the tip of the flame and the surface of the fuel (Figure 4.1). The flame length is an average because its instantaneous velocity is fluctuating in a random manner about the mean stream velocity due to turbulent flow.

Byram (1959) and others (Anderson 1966, Rothermel and Anderson 1966, Van Wagner 1968, Thomas 1971, Sneeuwjagt and Frandsen 1977) have given empirical arguments for the relationship of flame length and fire intensity. The equation given by Byram has the best empirical evidence over the widest range of intensities (cf. Alexander 1982):

$$I = 259.83\,L^{2.174}$$
$$\text{or} \qquad L = (I/259.83)^{0.46} \qquad\qquad (4.2)$$

where I is in kW m^{-1} and L is in m. Crown fires require that one-half the mean canopy height be added to L.

Recently, Nelson and Adkins (1986) have found that in fires burning with small pine needle fuel consumption (*c.* 0.5 kg m^{-2}), flame length is a

Table 4.2. *Classification of frontal fire intensities and kinds of fires for the boreal forest (Van Wagner 1983)*

kW m^{-1}	Kind of fire
< 10	Smouldering fire in deep organic layer
100–800	Surface backfire
200–15 000	Surface headfire
8000–30 000	Crown fire with single front
up to 100 000	High-intensity spotting fires

constant 0.5 m regardless of intensity. They suggest that fires may be limited in flame length by the interaction of the boundary layer of air above the forest floor and the buoyancy generated by the fire. If the boundary layer inertial forces are greater than the buoyancy forces of the fire and mass flow of partially burned fuels, then the flame is contained within the boundary layer. Thus, under these conditions of small needle fuel consumption, equation 4.2 may not be appropriate.

Fire behavior and fire intensity

Fire intensities have been estimated in the boreal forest to reach as high as 100 000 kW m^{-1} (Kiil and Grigel 1969, Alexander 1982) although 30 000 kW m^{-1} is a more likely extreme. Van Wagner (1983) gives a classification (Table 4.2) of intensities associated with differences in fire behavior.

The sensitivity of intensity (*I*) can be determined by examining the variation in each of its variables. As we have seen in the preceding sections, the heat of combustion (*c*) is for our purposes a constant, fuel consumption varies about 10-fold while rate of spread (*r*) varies 100-fold (see also Van Wagner 1965). Consequently, the 1000-fold variation in fire intensity is most significantly influenced by the variation in the rate of spread. We have already considered in Chapter 3 why rate of spread should be so large in the boreal forest and now we see how it influences intensity. Later in this chapter we will explain how high intensities result in the preponderance of crown fires in the boreal forest.

By replacing heat of combustion and fuel weight by total fuel consumption (metric tons per hectare) in the fire intensity equation 4.1, a simple graph of intensity, rate of spread and fuel consumption can be

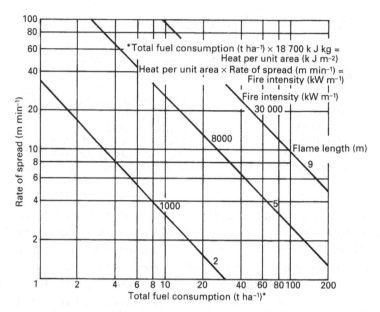

Figure 4.2. Fire characteristic chart which relates rate of spread, and total fuel consumed to fire intensity and flame length.

created (Figure 4.2). This fire characteristic chart can come in many forms, but is always derived from the basic relationship in Byram's equation:

$$I = r/60 \cdot H_A$$

where I is fire intensity (kW m^{-1}), r is rate of spread (m min^{-1}) and H_A is heat per unit area (kJ m^{-2}) which is H_A = heat of combustion total fuel consumed (kg m^{-2}). Since the heat of combustion can be considered a constant of c. 18 700 kJ kg^{-1} (cf. Van Wagner 1972b, Albini 1976) total fuel consumed and rate of spread become the only important variables.

The fire characteristic chart (Figure 4.2) shows a fundamental phenomenon in fire behavior: the inverse relationship between the rate of spread and the fuel consumed on fire intensity. This inverse relationship led Rothermel and Anderson (1966) to recognize two different types of fire behavior: wind driven (horizontally 'turbulent') fires and convectively driven (vertically 'turbulent') fires. Wind driven fires have high rates of spread and low fuel consumption while convectively driven fires have low rates of spread but large fuel consumption. Wind and topography clearly play important roles in determining these two types of behavior.

Different heat transfer mechanisms explain the differences in behavior between wind and convectively driven fires. Wind driven fires spread by tilting the flame and convectively heating the fuel in front. Slope steepness has the same effects as increasing wind speed. Convectively driven fires, on the other hand, have low horizontal spread but marked vertical turbulence and for this reason are often called fire storms by analogy to thunderstorms. Heat transfer is by radiation. Convectively driven fires must create enough heat release per unit area to produce buoyancy effects which will overcome the horizontal wind (Byram 1966).

Wind and convective fires have clear implications not only for vegetation mortality, which we will consider in this chapter, but also for the amount of duff consumed (Chapter 5). Wind driven fires are more common than convectively driven fires because of the special condition required for the latter. Convectively driven fires, although still not well understood, seem to require an unstable atmospheric lapse rate to maximize buoyancy and a large number of point ignitions as occurs in extreme spot fires (Byram 1954, Countryman 1964, Byram and Martin 1970).

Fire intensity and crown fires

By definition the high intensity of boreal forest fires means that flames extend into and ignite the tree crowns. In this section a simple physical theory of crown fire initiation and spread (cf. Van Wagner 1977a) is used to understand some of the reasons why the boreal forest experiences a predominance of crown fires.

The fuel distribution in most forests (cf. Merrill and Alexander 1987) consist of a duff layer (ground fuel) corresponding approximately to the H and F layer of the soil, a litter layer (surface fuel) of leaves, needles, herbaceous vegetation, low and medium shrubs, tree seedlings, stumps and downed-dead tree boles and branches, and a crown layer (crown fuel) of standing and supported forest combustibles not in direct contact with the ground. Crown layer includes live foliage and small branches in the canopy. Ladder fuels are often recognized since they provide vertical fuel continuity between the surface and the crown fuels. Ladder fuels consist of intermediate sized trees, saplings, arboreal lichens, live or dead lower branches and bark flakes.

The flaming fire front involves the layer between the litter and crown. The duff and larger fallen tree boles usually burn after the passage of the flaming front by glowing combustion. Standing live tree boles do not burn. Fires are described by the fuel layers in which they are burning as either surface-ground fires or crown fires. Three types of crown fires can be recognized (cf. Van Wagner 1977a). Passive or intermittent crown fires do not burn continuously in the tree canopies but burn into the crown then drop back to the surface fuels. Consequently, trees adjacent to each other can suffer quite different fates. Active or dependent crown fires form a wall of flames from the surface into and above the tree canopy but depend on surface fire intensities, continuous and moderate canopy bulk densities. Independent crown fires, as the name suggests, burn independent of the surface fuel layer. This is a rare type and can exist for only short periods of time under certain very unusual heat transfer conditions. In the upland boreal forest, active crown fires are very common in closed canopied forest while passive crown fires are common in the open canopied lichen woodlands.

These three types of crown fires are the result of three limiting factors (Van Wagner 1977a): (1) a critical surface intensity which ignites the crown fuel, (2) a critical horizontal crown heat flux which results from the burning of the tree crowns but may be supplemented by unburned buoyant gaseous fuels from the surface fire, and (3) a critical rate of spread below which the fire in the canopy cannot continue.

The critical surface fuel intensity (I_0) to ignite the tree crown requires knowledge of the ignition temperature of the crown and the vertical temperature profile resulting from different surface fire intensities. The heat of ignition (kJ kg^{-1}) of the canopy tree fuel is the energy required to heat the fuel to 100 °C so water will evaporate (latent heat) and then to 300 °C so escaping gases will start to ignite (see Chapter 5 for complete derivation) which is given by:

$$h = 460 + 26m \qquad (4.3)$$

where h is the heat of ignition and m is the canopy fuel moisture expressed as a percentage dry weight.

Thomas (1963) derived the dimensional relation for the rise in temperature above ambient ΔT from the convective heat flux (I) of a flaming front and the height above the ground (z) to be:

$$\Delta T \propto I^{2/3}/z$$

or, solving for I

$$I = (\Delta T \cdot z)^{3/2} \qquad (4.4)$$

As it stands equation 4.4 does not take into account that the ΔT required to make I critical must be higher if the canopy foliage is moist (heat of ignition

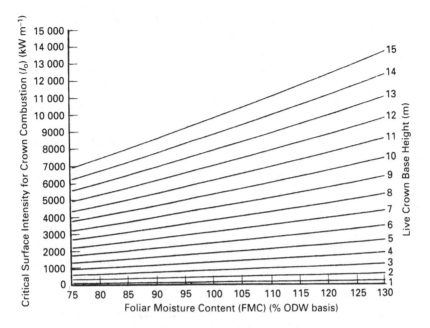

Figure 4.3. Critical surface fire intensity for crowning as a function of live crown base height and foliage moisture percentage (from Alexander 1988). Note that the empirical constant C in equation (4.5) is equal to 0.010 based on a minimum surface intensity for crowning of 2500 kW m^{-1}, 6 m crown base and 100% foliage moisture (cf. Van Wagner 1977a).

larger) and lower if the foliage is drier. Therefore, Van Wagner (1977a) combined ΔT and h into a term $(\Delta Th/h_0)$ which defines the temperature rise required for ignition. The heat of combustion h_0 is simply a mathematically arbitrary value which ensures that the ratio $h/h_0 = 1$ when heat of ignition plays no role in equation 4.4.

Substituting h/h_0 into equation 4.4 gives the critical fire intensity required to initiate crowning:

$$I_0 = (Czh)^{3/2} \qquad (4.5)$$

where $C = \Delta T/h_0$ is an empirically estimated constant. Notice that in this derivation I_0 is dependent only on the crown height and foliage moisture and not on crown geometry or bulk density. Figure 4.3 illustrates equation 4.5 in graphic form.

Boreal forest trees have variable crown base heights at maturity which depend on the quality of the site, the stand density and the species' self-pruning habit. Jack pine, red pine, white pine and aspen have elevated crowns and shed their lower and dead branches which leads to a separation of their crowns from the ground. Crown heights of 10 m are typical in mature jack pine and higher in red and white pine. Black spruce, white spruce and balsam fir have lower crown bases and often do not shed their lower dead branches. Spruce and fir, because of the presence of lower branches with often abundant arboreal lichens on them, have crown bases 5 m or less from the ground.

The crown architecture of conifers is deeper (several layers), has needle-shaped leaves, and can use smaller branches to support their leaves. Deciduous trees, in general, have single or two-layered canopies, larger leaves, and require larger branches to support their leaves. From the point of view of the amount of small diameter fuels, conifers are architecturally easier to burn than deciduous trees. However, as has been pointed out by Sprugel (1989) these 'flammable' traits of conifers can be understood as a means of light-harvesting which disperses direct beam radiation over many layers of leaves, thereby reducing light saturation energy wastage. Further, at low temperatures net photosynthesis is reduced more in light saturated leaves than in unsaturated leaves, clearly an advantage in a boreal climate.

The canopy moisture of the boreal forest is lower than the deciduous forest canopy and also has a spring and summer decline. Deciduous leaves have a foliar moisture content of 150–200% dry weight while conifer needles have a moisture content of < 100–150%. This difference seems to play a significant role in the heat required to ignite the canopy foliage. Figure 4.4 shows graphically the difference in heat of ignition by increasing foliar moisture. Also, the canopy foliar moisture in the conifers of the boreal forest is lower in spring and early summer (Van Wagner 1967, Fuglem and Murphy 1980, Springer and Van Wagner 1984, Chrosciewicz 1986) than in deciduous forest trees. This dip (Figure 4.5) is believed to be responsible for the susceptibility of conifer stands to crown fires early in the fire season (Van Wagner 1974). The spring decline in moisture does not vary in timing from year to year and consequently is not a weather induced phenomenon; rather, it is related to physiological changes which increase the starch content of the needles, resulting in the decrease in moisture (Pharis 1967, Little 1970).

In the boreal forest, live crown bases generally vary from the forest floor up to 10–15 m and have foliar moisture from 75 to 130%. Using these ranges, Figure 4.3 shows that live crown bases are much more important in controlling the critical crowning fire intensity than the foliar moisture. If we divide the major trees in the boreal forest into three classes based on their

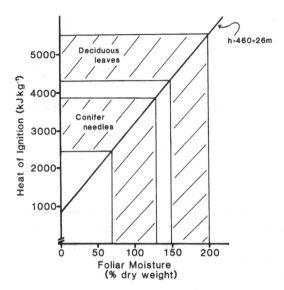

Figure 4.4. The importance of the latent heat in the heat of ignition of a canopy can be seen in the increase in heat required to ignite a canopy at higher foliar moisture.

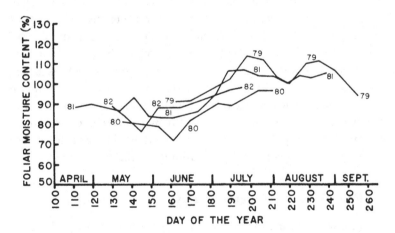

Figure 4.5. Seasonal trend in foliage moisture contents for old black spruce needles at Kapuskasing, Ontario over a 4 year period (cf. Springer and Van Wagner 1984). For comparison, deciduous leaves start at 300% foliage moisture in May and decline to a constant 150% moisture by mid-June.

mature tree heights on good sites, spruce and fir should require the lowest intensity to start crowning, perhaps as low as 1500 kW m^{-1}; the pines require the next highest intensities, perhaps as low as 2500 kW m^{-1}; and aspen, because of its high foliar moisture and crown height, has the highest intensities.

As flames extend from the ground and engulf the crown, the flaming front must move in order to find new fuel. Clearly, crown fires must have some critical rate of spread in order to maintain this flow of fuel. This minimum rate of spread (r_c) is

$$r_c = E_o/cd$$

where E_0 is a critical net horizontal heat flux, d is the bulk density of the crown space (i.e. not just the tree crowns) and c is the ignition energy of the fuel. The component (E_0/c) is called the mass flow and has been assumed to be a value around 3.0 kg/m^2·min (cf. Van Wagner 1977a) for solid crown flames. Consequently the rate of spread is

$$r_c = 3/d \tag{4.6}$$

Figure 4.6 gives this relationship between the minimum rate of spread r_c and the range of crown space bulk densities usually found in the boreal forest.

Most boreal forest stands have canopy bulk densities higher than 0.20 kg m^{-3} and consequently (Figure 4.6) require a relatively low rate of spread (15 m min^{-1}) to sustain a crown fire. Below crown bulk densities of 0.05 kg m^{-3} active crown fires are almost impossible because of the high rate of spread required. Clearly, open canopied stands such as lichen woodlands require much higher rates of spread in order to maintain an active crown fire than do closed canopied stands. In fact, the rates of spread in lichen woodlands do seem to be generally higher than in closed boreal forest stands. This may be because of the more rapid drying of the surface fuel, their open stand geometry and continuity of the surface fuels.

Fire intensity and tree mortality

Fire intensity may kill or damage trees primarily by girdling the stem or by scorching or consuming the leaves, needles and buds. Because fires and trees are variable, tree mortality is usually given as probabilities which incorporate a series of steps coupling the fire to the individual tree and then the local population. Peterson and Ryan (1986) give a flow-chart of some of the factors and couplings between fire behavior and the fire

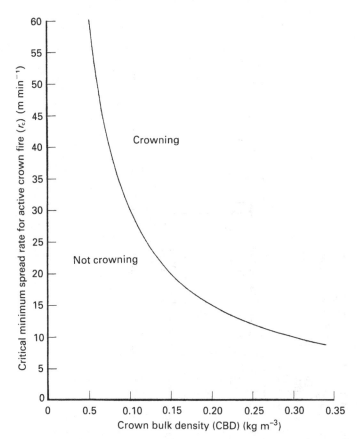

Figure 4.6. The relationships of crown spruce bulk density to rate of spread following equation 4.6 (from Alexander 1988).

effects on tree mortality (Figure 4.7). Let us now consider in some detail the mechanisms which may be responsible for crown kill, cambial kill and the resulting tree mortality in more detail.

Crown kill

The crowns of boreal trees are killed by fire when heated above a lethal temperature of 60–70 °C (Kayll 1968, Methven 1971). The derivation of the height of the flame at which this lethal temperature occurs (cf. Van Wagner 1973a) begins with Thomas's (see equation 4.4) relation for rise in temperature ΔT with fire intensity (I) and height above the ground of

$$z = I^{2/3}/\Delta T \qquad (4.7)$$

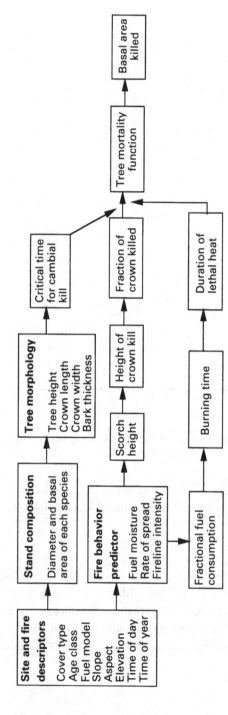

Figure 4.7. Some of the factors and couplings required to understand fire behavior effects on tree mortality (from Peterson and Ryan 1986).

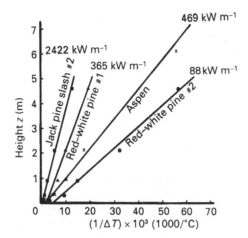

Figure 4.8. Data from four experimental fires showing the linear relationship between the height above ground (z) and the inverse of the maximum rise in temperature ($1/\Delta T$) in a fire convective plume as proposed in equation 4.7 (from Van Wagner 1975). The slopes of the lines reflect the intensities of the fires.

Figure 4.8 shows this relationship in four boreal forest fuels. With a light wind (u), the angle (A) of the plume is (cf. Thomas 1963)

$$\tan A \propto (bI/u^3)^{\frac{1}{2}} \tag{4.8}$$
$$\text{and } b = g/\rho_a c_a T_0$$

where b is a bouyancy term, g is the acceleration due to gravity, ρ_a is the density of air, c_a is the specific heat of air and T_0 is *absolute* temperature of the ambient air. The value of b is essentially constant over the range of ambient temperatures likely to occur during the fire season in the boreal forest.

The height of the plume at specific temperature rise is thus

$$z \propto \frac{I^{2/3} \sin A}{\Delta T} \tag{4.9}$$

where $\sin A \propto [bI/(bI + u^3)]^{\frac{1}{2}}$ from equation 4.8. The choice of a lethal temperature to replace ΔT depends on whether buds or foliage are to be considered, the season of the year, and, for buds, if they are shielded by the foliage. Using a general lethal temperature for boreal trees of 60 °C, empirical validation of the relationship in equation 4.9 is shown in Figure 4.9.

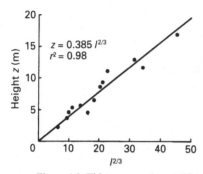

Figure 4.9. Thirteen experimental fires in eight red and white pine, two jack pine, one red oak and two red pine stands. The line is the expected relationship between scorch height (z) and fire-line intensity (I) under windless conditions in equation 4.9 (from Van Wagner 1973a).

Cambial kill

The bole of a tree can, as a first approximation, be considered a semi-infinite slab which is suddenly heated to a temperature T_f, and the bark having a thermal diffusivity of a. Then the time required to raise the temperature of the cambium inside the bark (T_c) to a lethal 60 °C may be calculated from (cf. Kreith 1965)

$$\frac{T_c - T_f}{T_b - T_f} = \mathrm{erf}\left(\frac{x}{2\sqrt{a(t)}}\right) \tag{4.10}$$

where T_b is the ambient bark temperature, x is bark thickness, t is time and erf is the Gaussian error function (values of this function can be found in most books of mathematical tables). If we assume as Peterson and Ryan (1986) do that $T_c = 60$ °C, $T_0 = 20$ °C, $T_f = 500$ °C and $a = 0.060$ cm² min⁻¹ (cf. Martin 1963), then

$$0.917 = \mathrm{erf}\left(\frac{x}{2\sqrt{0.06t}}\right)$$

Solving for t we obtain the time in minutes required to kill the cambium:

$$t = 2.9x^2 \tag{4.11}$$

This is the same scaling value that Hare (1965) found empirically from an assortment of barks in trees in the southern United States. Hare states that the time required to raise the cambium to the lethal level of 60 °C is directly proportional to the square of the bark thickness. Average bark thickness

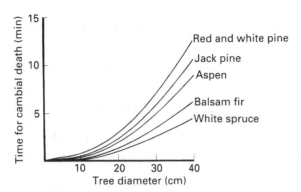

Figure 4.10. The time to cambial death as a function of tree diameter using equation 4.11.

increases with diameter (Martin 1963, Kayll 1963, Hare 1965). Consequently, besides species differences in bark thickness, diameter of the tree is of essential importance.

Observations in several different tree species (McConkey and Gedney 1951, Ferguson 1955, Ryan 1982, Peterson 1983) have shown that crown and cambial kill are positively related.

Tree mortality

Because active crown fires are so common in the boreal forest, complete tree mortality is usual. Only in passive crown fires and high intensity ground fires do trees generally survive. Jack, red and white pine survive often enough that their fire scars are useful in dating fires. This is because of their thicker bark (Figure 4.10) and generally greater crown base height. Pines have two further advantages in surviving fires. Their rapid height growth rate quickly elevates their crown base height and the larger diameter results in thicker bark (note increase in bark as function of diameter in Figure 4.10). Black and white spruce are easily killed by fire because of their thin bark (Figure 4.10) and lower crown base heights. Spruce also have more small branches and their buds are more exposed than pines. Aspens have very thin bark and thus are killed by even short periods of heating of the basal cambium.

In conclusion, high intensity fires are the rule in the boreal forest and most boreal conifers suffer high mortality at low fire intensities owing to their canopy architecture, low foliar moisture and thin bark.

Appendix: Mechanism for scaling fire intensity to flame length

The purpose of the next few paragraphs is to give a mechanism which might account for the allometric exponent of 0.46 in equation 4.2. The *empirical* scaling proposed in equation 4.2 does not give an explanation but merely states that for a change in unit flame length, fire intensity should increase by an exponent of 0.46. What follows is a model of the aerodynamics of a flame which gives a possible mechanism. The derivation is that of Fang (1969) and Nelson (1980).

Flame shape is determined by the rate at which air is entrained with the volatilized fuel and burns. The following assumptions are incorporated into the formulation: the flame can be considered a jet of hot gas being ejected into the air, the flow is fully turbulent, the rate of air entrainment into the buoyant turbulent jet is proportional to local *axial* velocity, the velocity is constant across the jet, ideal gas laws apply and mixing of volatile gases and air are in stoichiometric proportion. The flame is considered to be a buoyant turbulent jet whose length is determined by the combustion of the fuel. The top of the flame marks the point at which combustion and flaming stop. The following argument is for headfires in light wind. The notation is given in Figure 4.1. A more detailed derivation can be found in Fang (1969) or Nelson (1980).

The first equation invokes two ideas: a control volume and the conservation of mass. The control volume is the part of the flame (dx) in Figure 4.1 across whose surface the mass moves. The conservation of mass in this control volume is:

[The rate at which mass enters] = [The rate at which mass leaves the control volume]

and as an equation:

$$\frac{d(uy)}{dx} = \frac{\rho_a}{\rho_0} E'm \tag{A4.1}$$

where ρ_0 is the density of volatilized fuel and combustion products, ρ_a is the density of surrounding air, E' is a constant which describes the entrainment of ambient air, u is the local axial velocity in the flame and y is half the axial distance in the flame.

Now simplify by letting $A=(\rho_a/\rho_0)E$ where $E=2E'$ and $B=gD(\rho_a-\rho_0)/\rho_0 u_0^2$ where g is the acceleration due to gravity, D is flame width and u_0 is the velocity at the base of the flame. Further let $a=u'y'$ where $u'=b/a$, $y=a^2/b$ and $b=u'^2 y'$. Then the conservation equation in its dimensionless form becomes

$$da'/dx' = Au' = Ab/a \tag{A4.2}$$

$$db'/dx' = B \tag{A4.3}$$

Integrating A4.3 subject to $b=\cos\theta=y'$ and $x'=0$ gives

$$b = Bx' + \cos\theta \tag{A4.4}$$

Integrating A4.2 with A4.4 and using the boundary conditions $a=\cos\theta$ when $x'=0$ gives

$$a = [ABx'^2 + 2A\cos\theta x' + \cos^2\theta]^{\frac{1}{2}} \tag{A4.5}$$

The flow of a mass of air R into the flame tip per unit mass of unburned fuel is

$$2\rho_0 uy/\rho_0 u_0 D = u'y' = a' = R+1 \tag{A4.6}$$

Therefore if $x'=L/D$, where L is the flame length, then $(R+1)$ mass of air and combustion products flow through the flame for every unit mass entering at D.

The terms $2A\cos\theta x'$ and $\cos^2\theta$ are small enough to be ignored. Equation A4.5 can be added to A4.6.

$$\frac{2\rho_0 uy}{\rho_0 u_0 D} = (ABx'^2)^{\frac{1}{2}} = E\frac{\rho_a}{\rho_0}\left(\frac{gD(\rho_a-\rho_0)}{\rho_0 u_0^2}\right)\left(\frac{L^2}{D}\right)$$

$$L^2 = \frac{(R+1)^2\,\rho_0^2 u_0^2 D^2}{E\rho_a gD(\rho_a-\rho_0)} \tag{A4.7}$$

Now, to introduce fire intensity (equation 4.1) we must introduce into equation A4.7 r, m and c by assuming that the rate of combustion $(\rho_0 u_0 D)$ is equal to the arbitrary value $r/5$ times the mass burning rate per unit length of fire line.

Making these substitutions where $I=cmr$ gives

$$\rho_0 u_0 D = \left(\frac{R+5}{5}\right)rm = \left(\frac{R+5}{5}\right)\frac{I}{c}$$

Now,

$$L^2 = \frac{(R+1)^2\,(R+5)^2\,(I/c)^2}{5E\rho_a\,gD\,(\rho_a - \rho_0)}$$

$$L = \frac{R^2 + 6R + 5}{5\rho a\,H}\left(\frac{I_R}{Eg(1 - \rho_0/\rho_a)}\right)^{0.5}I^{0.5} \tag{A4.8}$$

where $I_R = I/D$. Thus $L \propto I^{0.5}$ is very close to Byram's empirical result of $I^{0.46}$ and the relationship of flame length to fire intensity may be explained by the behavior of the flame as a buoyant turbulent jet of hot gases drawing in cooler surrounding air as it burns.

5

Duff consumption

The forest floor of the upland boreal forest consists of organic matter resting on the mineral soil surface. The organic matter can be divided into two indistinct layers: litter and duff. The litter layer (L) consists of the loosely packed, largely unaltered dead remains of animals and plants usually recently cast. The duff has two layers: (1) an upper F layer consisting of litter which has recently begun to decompose but with the particles still recognizable as to their origins and (2) a lower H layer which is made of well-decomposed organic matter which can not be recognized as to its origins.

In the upland boreal forest the thickness of the forest floor increases with colder soil and air temperatures (Figure 5.1). This leads to an increase in forest floor depth northward on similar sites but also locally on sites with different heat budgets. Generally, dry, warm sites with canopies of jack pine and white spruce have thinner forest floors and cooler, wetter sites with canopies of black spruce have thicker forest floors (Johnson 1981). The increasing forest floor depth is primarily caused by an increasing duff layer.

As we have already discussed in Chapter 3, forest fires spread mostly by flaming combustion in the surface litter layer because its large surface area and loose packing allow rapid drying. The duff, on the other hand, burns mostly by glowing combustion during and after the flame front passage, although some volatiles are evolved and ignite in flames. The glowing combustion is a result of decay which has left as fuel the less easily decomposed lignin. Lignin burns by glowing combustion compared to the flaming combustion of cellulose (Rothermel 1976). Further, the compactness of duff leads to less aeration and a larger water holding capacity than litter. Thus more heat is required to evaporate the water before the duff can be heated to its ignition point. Duff with moistures below approximately 30% dry weight burns unassisted after ignition while duff with moisture contents above 140% generally cannot burn (cf. Wright

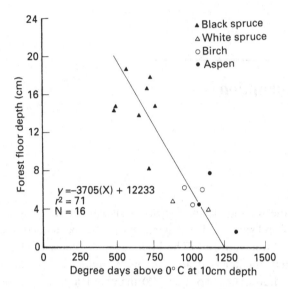

Figure 5.1. The decrease in forest floor (litter and duff) depth with increasing degree days above 0 °C at 10 cm into the soil in Alaska (cf. Van Cleve *et al.* 1981).

1932, Shearer 1975, Norum 1977, Artley *et al.* 1978, Sandberg 1980).

This chapter will consider the drying process of *upland* boreal forest duff, how fire consumes duff, and the coupling of duff consumption to plant mortality and seedbed preparation.

Duff drying processes

Duff has a porosity of 60–90% of free air and the movement of moisture in it during drying is generally considered to be by evaporation, internal particle diffusion and bound water loss (Simard 1968). Duff can be defined as the organic particles and the spaces between them. Both the particles and the spaces are subject to the movement of water by latent heat, sensible heat and mass transfer. The driving forces of these processes are temperature and humidity in the particles and spaces.

Fosberg (1975) gives a description of the drying processes in conifer forest litter and duff (Figure 5.2). The variables are the moisture content, heat and temperature of the organic particles, and the spaces between the particles. The processes coupling the particles and the spaces are shown as horizontal arrows. These consist of time-lagged transfers of heat and moisture between the particles and the spaces. The rate of transfer into and

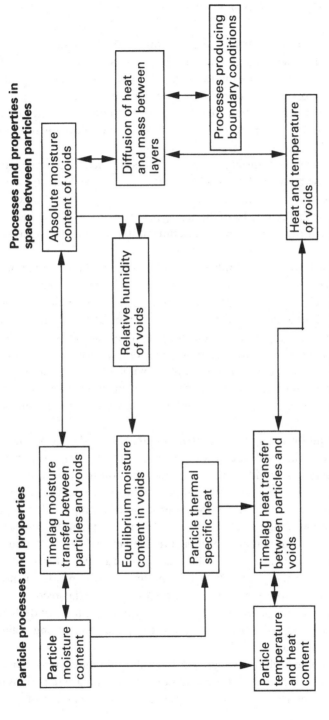

Particle processes and properties

Processes and properties in space between particles

Figure 5.2. Routes of heat and mass transfer processes in conifer forest litter and duff (cf. Fosberg 1975).

out of the organic particles depends on (1) the boundary conditions of temperature and relative humidity in the spaces, i.e. how the adjacent environment influences the forest floor environment, (2) the gradient of heat and moisture between the particles and the spaces, i.e. the difference between the heat and moisture of the particles compared with the spaces, and (3) the surface properties of the particles which slow down or speed up the uptake or loss of heat and water vapor.

Notice that the transfers within the spaces between the particles are rapid while those between the particles and the spaces have time-lags before the particles reach an equilibrium moisture content. These time-lags are important characteristics of the drying processes in all types of fuels and are determined by the size, shape, interior diffusivity and specific heat of the particles.

The drying process of duff is usually characterized by the relationship of the realized moisture relative to its equilibrium moisture. The equilibrium moisture content of duff is the value that the duff moisture would approach if exposed to a constant environment of temperature and humidity for a very long time. Equilibrium moisture content is a steady-state condition in which the gain and loss of moisture from the duff is the same. The equilibrium moisture content is unfortunately not a unique number but differs depending on whether the fuel was drying or wetting compared with the environment. Here we will always use the drying equilibrium moisture content.

The equilibrium moisture content (Figure 5.3) divides the bound and free moisture in the duff. Bound water is that water exerting a vapor pressure less than that of liquid water at the same temperature. This water is usually held to the duff particles by hydrogen bonds and is not available to the drying process. Free water is the water removed during the drying process. For example in Figure 5.3, the mixed red and white pine duff at 60% relative humidity has an equilibrium moisture content of 11%. If the duff has a moisture content in a forest of 45% its moisture available for drying is 34% $(45-11)$, not the total 45%. Under natural conditions, duff is probably never at equilibrium moisture. Figure 5.4 shows a 35 day record of the free moisture content of red pine duff. The constant wetting and drying of the duff is clearly the rule. Notice, however, that the drying follows an exponential rate.

The course of duff drying towards equilibrium moisture is divided into two phases (Figure 5.5): the constant rate and falling rate periods (Perry, 1963). The constant rate period never lasts very long in duff or litter (Van

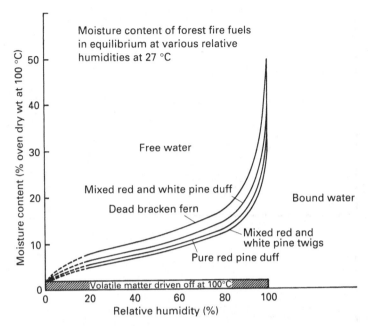

Figure 5.3. The equilibrium moisture content of selected duff types at different relative humidities (cf. Wright 1932).

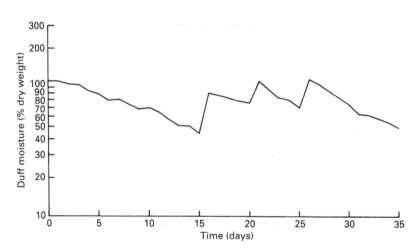

Figure 5.4. A 35 day record in June 1965 of the moisture content as percentage dry weight in red pine duff layer weighing 3.3 kg m^{-2}. Note that moisture is on a log scale to show the exponential trend of drying (cf. Van Wagner 1970).

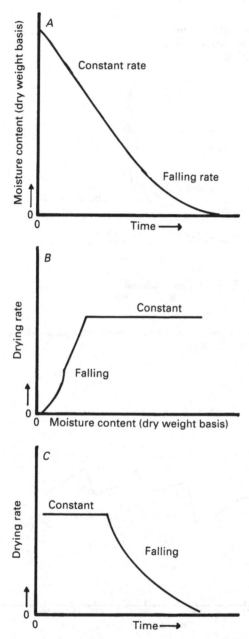

Figure 5.5. Duff drying periods. *A*. Moisture content versus time. *B*. Drying rate versus moisture content. *C*. Drying rate versus time.

Wagner 1979) and is the result of evaporation type processes. During this phase, the duff surface is always saturated and the movement of water to the surface of the duff particles is rapid enough to maintain saturation. Evaporation from a water surface is independent of the properties of the particles in the water. The only control besides the vapor pressure and temperature is the wind, as it affects the depth of the boundary layer and thus the diffusion resistance. Within the boundary layer only molecular diffusion takes place and not the turbulent transport which occurs above it.

The rate of moisture loss during the constant rate period is the same as a mass transfer process with the only resistance being related to the boundary layer:

$$\frac{dw}{dt} = \frac{hA\Delta T^{\circ}}{\lambda}$$

where w is the moisture content, t is time, A is area, ΔT° is the temperature difference between and duff particles and the spaces between the particles, λ is latent heat of vaporization and h is the heat transfer coefficient from the spaces to the duff particles.

Solving for time during the constant rate period (t_c):

$$t_c = \frac{w - w_c}{R_c A} \tag{5.1}$$

where R_c is the drying rate per unit area, A is the exposed surface area, and $(w - w_c)$ is the difference between the initial moisture content and the critical moisture content (see Figure 5.5).

The falling rate period is the most important of the duff drying processes because the duff's moisture content is now low enough that heat from a fire could evaporate the water and ignite the duff. The falling rate is a result of a shift from the completely saturated particle surface of the constant rate to a less and less saturated surface. The drying mechanism shifts from evaporation where wind is important to internal diffusion where wind is not important. The falling rate period ends when the duff is at equilibrium moisture with only bound water left.

During the falling rate period, the duff has (Figure 5.6): (1) an exponential drying rate under constant weather conditions; (2) a log drying rate which is inversely proportional to the dry weight of the duff layer; and (3) a log drying rate which is directly proportional to the relative humidity (Van Wagner 1970, 1979). These empirical observations are consistent with a falling rate process:

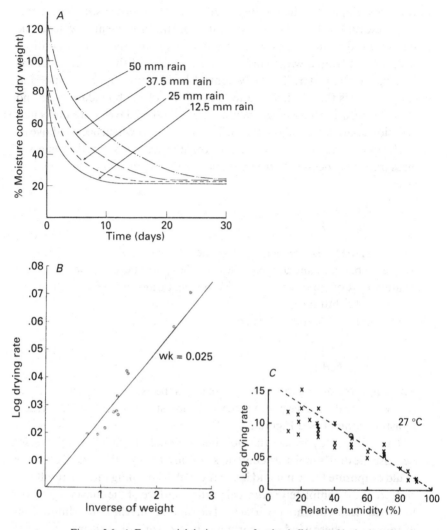

Figure 5.6. *A*. Exponential drying curves for the duff layer of jack pine after different amounts of precipitation (cf. Stocks 1970). *B*. The inverse relationship between the log drying rate and weight per unit area of red pine duff (cf. Van Wagner 1970). *C*. The log drying rate versus relative humidity for jack pine litter (cf. Van Wagner 1979).

$$\frac{dw}{dt} = k(w - \text{EMC})$$

where EMC is the equilibrium moisture content. Solving for time during the fall rate period (t_f) gives

$$t_f = \frac{1}{k} \ln\left(\frac{w_c - \text{EMC}}{w - \text{EMC}}\right)$$

or

$$\frac{w_c - \text{EMC}}{w - \text{EMC}} = \exp\left(-k_f\right) \tag{5.2}$$

where k is the drying rate.

Since the falling rate approaches the equilibrium moisture content asymptotically, the rate (k) is defined as the time required to reach 36.8% of the moisture. The time-lag of the duff is thus equal to the inverse ($1/k$) of the rate.

Most duff is found to follow the exponential drying rate given in equation 5.2 (e.g. Nelson 1969; Van Wagner 1969a, 1979, 1982); however, some forms of duff dry linearly with time, most notably lichens (Mutch and Gastineau 1970).

Duff burning process

Duff is burned by driving off the moisture and raising the duff to its ignition temperature by the heat from flaming or glowing combustion. Models of duff burning follow two approaches: regression models which relate the observed duff removed to measured variables, usually duff moisture (Brown *et al.* 1985), and heat budget models which propose heat transfer mechanisms (Van Wagner 1972a, Albini 1975). Both approaches require simplifying assumptions in order to make adequate predictions. The most important assumptions concern the depth, bulk density and moisture distribution of the duff. If the duff depth is shallow (approximately 8 cm) and bulk density around 5 kg m^{-2}, and moisture is not significantly different at different layers, then once the surface layer of duff is ignited, it could supply enough heat to ignite the lower levels.

The heat budget model which follows was developed by Van Wagner (1972a). The exchange of radiation between the flame and the duff depends on their emitting and absorbing characteristics and the view the radiation flux from each has of the other. If we assume that the flaming front provides the only heat for ignition and the heat flux is by radiate transfer as described by the Stefan–Boltzmann law:

$$E_f = \epsilon_f \sigma T_f^4$$

where ϵ_f is the emissivity of the flame (the ratio of the radiative flux emitted by a body to that emitted by a black body at the same temperature. A black body is an object which radiates at σT_f^4), σ is the Stefan–Boltzmann constant and T_f is the temperature of the flame in degrees Kelvin. The radiative flux (q_f) from the flame to the duff is

$$q_f = a_d F_f E_f A_f$$

where a_d is the absorptivity of the duff, F_f is the view the fire has of the duff and A_f is the radiate surface of the flame. The radiate flux (q_d) from the duff to the flame is similarly

$$q_d = a_f F_d E_d A_d$$

The net interchange from the flame to the duff, assuming the view and area terms are the same and therefore can be ignored, is

$$q_{net} = q_f - q_d$$

$$q_{net} = a_d E_f - a_f E_d$$

$$= a_d \epsilon_f \sigma T_f^4 - a_f \epsilon_d \sigma T_d^4 \qquad (5.3)$$

Since by Kirchhoff's law we can write equation 5.3 in terms of either emissivity or absorptivity (whichever is easier to measure), the radiative flux can be written:

$$q_{net} = a_d \epsilon_f \sigma (T_f^4 - T_d^4) \qquad (5.4)$$

or

$$q_{net} = a_f \epsilon_d \sigma (T_f^4 - T_d^4)$$

Choosing $a_d \epsilon_f$, assume that the duff absorptivity is 1, its temperature is 20 °C and the flame's temperature is 800 °C (cf. Van Wagner 1972a). Substituting these values into equation 5.4 gives $T_f = 1073$ °K and $T_d = 293$ °K, and putting the 10^{-12} of the constant $\sigma = 5.669 \times 10^{-12}$ W cm^{-2} K^{-4} into the denominator of the temperature terms:

$$q_{net} = 5.669 \epsilon_f \left[\left(\frac{1073}{1000} \right)^4 - \left(\frac{293}{1000} \right)^4 \right]$$

$$= 7.473 \epsilon_f$$

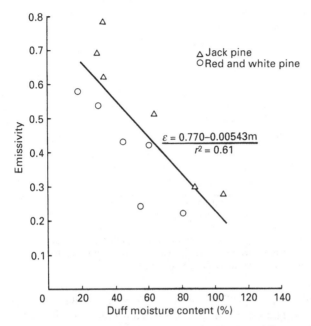

Figure 5.7. Relationship of decreasing downward flame emissivity (ϵ) with increasing duff moisture content (cf. Van Wagner 1972a).

Thus the net radiation flux from the flame to the duff is proportional to the emissivity of the flame. Emissivity (ϵ_f) is the efficiency of the heat flux.

The moisture content of the duff could significantly affect the emissivity of the flame by changing the amount of water vapor, carbon dioxide and soot particles in it. In fact, Van Wagner (1972a) has shown that there may be an inverse relationship between emissivity and duff moisture for jack, red and white pine duff (Figure 5.7).

The heat required to raise a unit mass of duff to ignition temperature must: (1) heat the water in the fuel to the boiling point: $mC_w(100° - T_0)$ where T_0 is the fuel temperature, C_w is the specific heat of water and m is the percentage moisture based on dry weight; (2) vaporize the water: lm where l is the latent heat of vaporization, 2250 kJ kg^{-1}; (3) heat the dry fuel to ignition: $C_f(T_{ig} - T_0)$ where C_f is the specific heat of duff, 1.47 J g^{-1} °C, T_{ig} is the ignition temperature of 300 °C; and (4) give up the heat of desorption of water in the fuel: 50.23 J g^{-1}.

Therefore, the heat required for ignition at 20 °C is

$$h = mC_w(100 - 20) + lm + C_f(300 - 20) + 50.23$$
$$= 2585m + 460$$

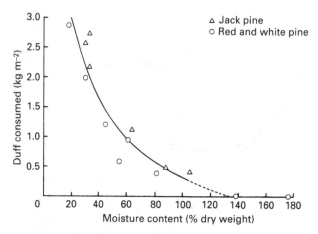

Figure 5.8. The line is the duff consumption using equation 5.5 and the emissivity relationship given in Figure 5.7. The dots show empirical duff consumption and moisture values (Van Wagner 1972a).

The actual duff burning is dependent on a heat flux high enough to cause the duff to burn (q'_{net}). This critical heat flux is written as an intensity just as was done in equation 4.1, $q'_{net} = h\, dM/dt$.

Now, solving for the weight of the duff consumed (d) gives

$$d = \frac{q'_{net}t_r}{h}$$

where t_r is the resident time of the flame, i.e. horizontal flame thickness divided by rate of spread. By substitution for h and q'_{net}, the relationship of duff consumed as a function of the emissivity and the duff moisture gives

$$d = \frac{7.473\epsilon_f \cdot 60}{2585m + 460} \tag{5.5}$$

where 60 is $t = 60s$. Figure 5.8 shows graphically the relationship between the duff consumed to duff moistures and emissivities.

Clearly, this model is a simple outline of the possible mechanisms involved in duff burning. It must be used with many reservations, particularly about the downward emissivity values and the rather contrived argument that the flame combustion radiation, but not the glowing combustion radiation, adequately describes the heat flux which ignites the duff. These are clearly areas of future refinement. However, at present the question is: what insight does this heat budget offer for the ecological effects of duff burning? The most obvious implication is that small changes in duff moisture can lead to large changes in the duff consumed, particularly below about 50% moisture. Within a single 10×10 m plot one can expect small areas, say 20 cm^2 samples, of duff to vary between themselves as much as their mean duff moisture. This would suggest that except at very low duff moistures, large amounts of variation in duff consumption should be the rule.

Duff consumption, plant mortality and seedbed preparation

Duff burn-out causes mortality in plants whose roots and seeds occupy the duff. Since the amount of duff removed is a function of soil moisture, low duff moisture will cause the duff to be consumed to a greater depth. Mortality of plants, particularly herbs and shrubs, from duff burn-out is thus primarily a result of the rooting depth of the underground organs. This has led to numerous classifications (e.g. Table 5.1) based on rooting depth and fire survival (e.g. Ahlgren 1960, McLean 1969, Purdie 1977a,b, Flinn and Wein 1977). Plants with rooting and vegetative organs in the mineral soil have the best possibility of surviving (e.g. *Vaccinium angustifolium*, *Cornus canadensis*), while those with roots and rhizomes in the duff or simply resting on the duff surface are most often killed (e.g. *Arctostaphylos uva-ursi*, *Linnaea borealis*, most mosses and all lichens). Species having their vegetative organs in the duff are very dependent on the pattern of the duff burn-out. As yet we know little of this pattern and its causes. The only obvious pattern is that extensive burn-out tends to occur around the base of boreal trees owing to the drier duff and accumulation of debris (often from squirrel middens).

Actual studies of plant mortality by duff burn-out are few. Most confirm that the number of above ground stems after a fire are related in general to the number of stems before the fire, the duff thickness and moisture (Ahlgren and Ahlgren 1960, Calmes and Zasada 1982, Johnston and Woodard 1985, Noste and Bushey 1987). Although a large percentage of boreal species are capable of some form of vegetative reproduction, no obvious pattern exists between survival after fire, rooting depth and type of vegetative reproduction method.

Buried viable seeds could also provide a source of propagules to recolonize burns if they were buried deep enough in the duff to survive the duff's consumption. However, buried viable seeds decrease rapidly with forest floor depth (Kellman 1970, Strickler and Edgerton 1976, Moore and Wein 1977, Kramer and Johnson 1987). In studies of buried seeds before

Table 5.1. *Mean depth of underground regenerative tissue of boreal herbs and shrubs. The surface of the duff (F) layer is the zero point of reference (from Flinn and Wein 1977)*

Species	Mean depth (cm)
Species usually found in litter layer	
Gaultheria procumbens	2
Gaultheria hispidula	2
Trientalis borealis	2
Epilobium angustifolium	2
Maianthemum canadense	1
Coptis trifolia	Top
Species usually found in the F and H layer	
Aralia hispida	0
Aster acuminatus	Top
Uvularia sessilifolia	2
Dennstaedtia punctilobula	2
Medeola virginiana	3
Species usually found in mineral soil	
Anaphilis margaritacea	3
Aralia nudicaulis	6
Rubus strigosus	6
Lycopodium obscurum	6
Vaccinium angustifolium	6
Cornus canadensis	8
Vaccinium myrtilloides	8
Pteridium aquilinum	9
Rubus canadensis	9
Kalmia angustifolia	9
Osmunda claytoniana	11
Viburnum cassinoides	14

and after fires, Archibold (1979) and Moore and Wein (1977) have found significant reductions in the numbers of seeds. Although the relationships between duff consumption and buried viable seed survival are not well understood at this time, clearly, seeds shed shortly after the preceding fire, if they remain viable, would be buried the deepest in the duff and thus have the best chance of surviving the next fire. Unlike chaparral vegetation, few boreal species are presently known which require heat to induce germination. Conifer tree seeds are not viable in the soil for more than a few years (Thomas and Wein 1985b). It is tempting to speculate that the low buried viable seed populations, usually less than 500 m^{-2} (cf. Johnson 1975, Archibold 1979, 1980, 1984), are related in part to the large amount of duff

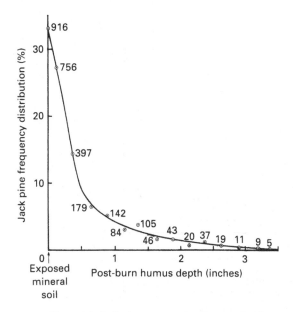

Figure 5.9. Jack pine regeneration 2 years after fire as a function of post-burn duff depth. The numbers of seedlings counted are indicated next to the plotted points. These data were collected from 11 burns on silty-fine sand and fine sand (cf. Chrosciewicz 1974).

consumed and hence the serious limitation of buried viable seeds as a regeneration strategy after fire.

The amount of recruitment by most boreal trees is strongly related to the amount of duff consumed and mineral soil exposed. Chrosciewicz (1974, 1976) and Weber *et al.* (1987) show for black spruce and jack pine that with increased exposed mineral soil the number of seedlings and their growth rates increase (Figure 5.9). The effect seems to be a result of a better germination substrate and early establishment environment. The exact nature of this advantage whether of reduced competition, more available nutrients or better soil–seed contact is again not known. Zasada *et al.* (1983, 1987) in a series of prescribed burns in black spruce stands found that most tree and shrub regeneration was in areas of large duff removal. Further, in planted seedlings of spruce, aspen, alder and willow, they found that spruce survived better than deciduous species depending on the depths of burn but had slower height growth. Height growth, but not seedling survival, was affected by depth of burn in all species.

Thomas and Wein (1985a) give a diagram (Figure 5.10) which

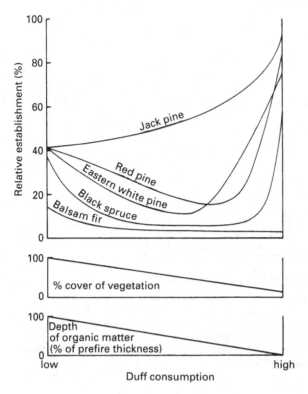

Figure 5.10. A hypothetical relationship between relative establishment of five conifers to duff consumption and vegetation shading (modified from Thomas and Wein 1985a).

summarizes their and 19 other studies of duff consumption on post-fire establishment. The diagram is hypothetical and assumes that along the gradient of duff consumption seed input is constant, climate is the same, ground cover is reduced linearly with duff consumption, treatments (artificial shading, etc.) simulate natural conditions, and differences in recruitment time do not affect early survival (e.g. serotinous species may have seed which germinate in the same season as the fire, while non-serotinous species may seed into the burn only in the next seed year).

All species except balsam fir recruit best on exposed mineral soil. Jack pine shows a decline with more duff remaining after fire (see also Figure 5.9). Red and white pine, which have limited ranges in the boreal forest, and black spruce have bimodal recruitment with highest establishment on both exposed mineral soil and no duff removed. This bimodal recruitment suggests an inverse interaction between the amount of exposed mineral soil and vegetation shading. Balsam fir has a distinct pattern when compared with other conifers. It does not recruit well on any surface but does slightly better on shaded and no duff removed surfaces than on any other combination.

In this chapter, I have spent some time explaining duff drying since the principles developed apply to all fuels. The importance of weather as it affects the duff moisture again presents itself as the key variable in determining duff consumption. However, duff is not consumed evenly and the impact of this on vegetation recovery has not been well researched. The possibility for experimental manipulation studies seems very inviting.

6

Fire history and landscape pattern

Up to this point, we have examined three fire behavior processes and effects which operate at an intimate level. We have considered forest fires in heat transfer terms and the effects on plants in terms of heat absorbed and combustion. However, it is also useful to step back and consider the fire's frequency and its implications for the tree population's hazard of death by fire. This landscape view has been of primary interest to ecologists (e.g. Knight 1987).

The boreal forest is a mosaic of stands each having different times since they last burned. Heinselman (1973) has called this landscape pattern the **stand origin map** (Figure 6.1). Most fire history studies are implicitly based on this map. The mosaic in the stand origin map is the result of a complex pattern of overlapping past fires. Only the most recent fires will appear on the stand origin map in their complete form, with all other fires being in some part obliterated by subsequent fires. The origin and meaning of this age mosaic is of central interest to any understanding of vegetation and fire dynamics in the boreal forest.

Two approaches can be used to understand the stand origin map. The **fire frequency** approach calculates the survivorship of stands in a landscape. The **fire pattern** approach is concerned with the age pattern of adjacent stands and the spatial autocorrelation in the stand origin map.

Fire frequency from the stand origin map

Fire frequency can be understood by the following examples used by Van Wagner (1978). Imagine a checkerboard which consists of a collection of 1000 squares (stands). Each stand ages each year until it is burned, which then sets it back to age zero. If all stands were burned as they

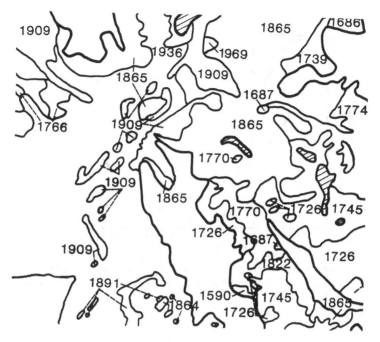

Figure 6.1. An example of a stand origin map.

reached 100 years, a rectangular distribution of stand ages would result after a finite period of time (Figure 6.2). If, instead of burning all stands as they reached 100 years, a constant percentage of the number in each age class were burned, a negative exponential distribution of stand ages would result (Figure 6.2). If, however, the burning always occurred in older stands, as opposed to younger stands, then a Weibull age distribution would result (Figure 6.2). The important point in these examples is that the manner in which the stands were chosen to burn results in a specific fire frequency distribution.

Fire history models

Two parametric models, the negative exponential and Weibull, have been used in fire frequency studies (Johnson and Van Wagner 1985). These models, as in all statistical distributions, can be presented in three forms, each of which is related explicitly to the others (Figure 6.3). The cumulative fire interval distribution for Weibull and the negative exponential respectively are:

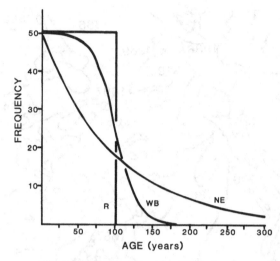

Figure 6.2. The time-since-fire distribution for a rectangular (R), negative exponential (NE) and Weibull (WB) distribution. All distributions have the same average fire frequency.

$$F(t) = 1 - \exp[-(t/b)^c] \qquad t > 0 \qquad\qquad\qquad \text{Weibull}$$
$$\qquad\qquad\qquad\qquad\qquad\qquad\qquad b > 0 \qquad\qquad\qquad\qquad\qquad (6.1)$$
$$F(t) = 1 - \exp[-t/b] \qquad c \geqslant 1 \quad \text{negative exponential}$$

where t is time, b and c are the scale and shape parameters: b is the fire recurrence in years or the fire interval which will be exceeded 36.79% of the time; c controls the shape of the distribution, as can be seen in Figure 6.3, and is dimensionless. $F(t)$ is the frequency of having fires with intervals less than age t, in other words it gives the cumulative mortality from fire.

The probability density fire interval distribution is

$$f(t) = \frac{\mathrm{d}F(t)}{\mathrm{d}t} = \frac{ct^{c-1}}{b^c}\exp\left[-\left(\frac{t}{b}\right)^c\right] \qquad \text{Weibull}$$

$$\qquad\qquad\qquad\qquad\qquad\qquad\qquad\qquad\qquad\qquad\qquad (6.2)$$

$$f(t) = \frac{1}{b}\exp\left(-\frac{t}{b}\right) \qquad\qquad\qquad \text{negative exponential}$$

This distribution gives the frequency of having fires within the interval $t - 1$ to t and thus is the frequency of mortality in a specific age class.

The (cumulative) time-since-fire distribution $A(t)$ is the complement of $F(t)$ and consequently is the frequency of not having a fire up to t:

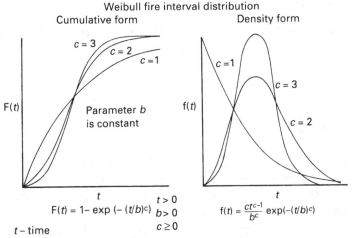

Weibull fire interval distribution

Cumulative form Density form

$$F(t) = 1 - \exp\left(-(t/b)^c\right)$$

$$t > 0$$
$$b > 0$$
$$c \geq 0$$

$$f(t) = \frac{ct^{c-1}}{b^c} \exp(-(t/b)^c)$$

t – time
b – fire recurrence (scale) parameter in years
c – shape parameter, dimensionless

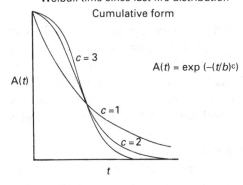

Weibull time since last fire distribution

Cumulative form

$$A(t) = \exp\left(-(t/b)^c\right)$$

Figure 6.3. The three distributions for the negative exponential ($c = 1$) and Weibull ($c \geq 1$) fire history model. Graphs show distributions with different shape parameters (c) but with a constant fire recurrence or scale parameter (b).

$$A(t) = 1 - F(t) = \exp\left[-\left(\frac{t}{b}\right)^c\right] \qquad \text{Weibull}$$

$$(6.3)$$

$$A(t) = \exp\left(-\frac{t}{b}\right) \qquad \text{negative exponential}$$

$A(t)$ is the chance of surviving up to t without a fire.

The manner in which the stands are chosen to burn is mathematically expressed in the model as follows. The fire interval distribution $f(t)$ can be written as the product of two distributions:

$$f(t) = \lambda(t) \cdot A(t) \cdot \qquad (6.4)$$

$f(t)$ is a product of the chance of surviving to age t ($A(t)$) and the hazard of burning $\lambda(t)$ during that interval. The hazard function is

$$\lambda(t) = \left(-\frac{1}{A(t)}\right)\left(\frac{dA(t)}{dt}\right) = \frac{ct^{c-1}}{b^c} \qquad \text{Weibull}$$

$$(6.5)$$

$$\lambda(t) = \frac{1}{b} \qquad \text{negative exponential}$$

The hazard of burning is clearly the instantaneous death or burning rate. From the point of view of fire history, $\lambda(t)$ contains the mechanisms that determine the fire frequency distributions; in a sense it defines the changing flammability of an average stand.

Another function which is sometimes useful in fire frequency studies is the expected number of fires in any time interval. This is called a renewal function (cf. Cox 1962):

$$H(t) = F(t) + \int_0^t H(t-x)\,dF(t)$$

For a negative exponential, the renewal function is $H(t) = 1/b$. This simple relationship results since the hazard of burning is constant, i.e. a Poisson process. The Weibull can be calculated only by a recursive procedure. A nomagram is given by Smith and Leadbetter (1963) for the Weibull renewal function. For both the negative exponential and Weibull, the renewal functions are functions of time with the slope dependent on the parameters.

Field methods in fire histories

The stand origin map (Figure 6.1) has, up to this point, proved to be the best method for collecting data to estimate the time-since-fire distribution ($A(t)$). Construction of the stand origin map is outlined in the following paragraphs.

Aerial photographs are used first to give the general outline of fire boundaries. Next, samples are distributed in the field to refine the boundaries. Within each sample all evidence is used to determine the time to the last fire; not only are fire scars used but growth release of surviving trees, reaction wood, trees recruited after the fire and dendrochronologically cross-dated fire killed trees. Care must be taken to not confuse evidence of other disturbances with fire (McBride 1983). Each fire date should be replicated first within the sample and then between samples. A useful device for showing the replication of dates both within and between sample plots is the fire event plot (Figure 6.4). However, the event plot cannot be used to adjust the accuracy of the fire dates as Arno and Sneck (1977) proposed. Madany *et al.* (1982) show that only dendrochronological cross-dating (Fritts 1976) can provide the accuracy of the dates.

When determining ages of trees, whole disks are preferred to wedges and increment cores. Whole disks allow the complete ring circuit to be traced at regular intervals in the count so as to detect locally missing or double rings. Ring anomalies are particularly common in certain species (e.g. black spruce), very old trees and severely injured trees. The easiest procedure for locating missing rings on disks is to mark two axes with a pencil. These axes should be away from the scar if one is present. The rings are then counted and marked at decade intervals along each axis and the ring of each appropriate decade traced around the disk to see if it matches the decade on the other axis. In wedges and increment cores, ring-counts along only the one axis cannot ensure accurate dates and the more laborious process of cross-dating must be used. Cross-dating is the procedure by which a master ring-width chronology is compared with an undated set of ring-widths to find a match. Rings in the undated tree then can have dates attached to them. The method when done properly is very accurate (Fritts 1976).

Analysis of fire histories

Once the stand origin map has been constructed, the area in each fire year on the map can be determined and expressed as a cumulative percentage, starting with the oldest year (Table 6.1). When plotted on either semi-log or Weibull probability[1] paper, the cumulative percentages give the graph of the time-since-fire distribution $A(t)$. Note that the proportion in each age class (column 3 in Table 6.1) is *not* the probability density fire interval distribution $f(t)$ since the probability density is a fire-to-fire interval and these data are last fire to present. The parameter of a negative exponential distribution (b) can be estimated easily by Maximum Likelihood as the sample mean. Estimating the two parameters of the Weibull is more difficult. They can be approximated by the graphic mean (King 1971), method of moments (Garcia 1981) and Maximum Likelihood (Harter and Moore 1965).

---

[1] Weibull paper can be obtained from Technical and Engineering Aids for Management, Box 25, Tamworth, New Hampshire 03886, USA

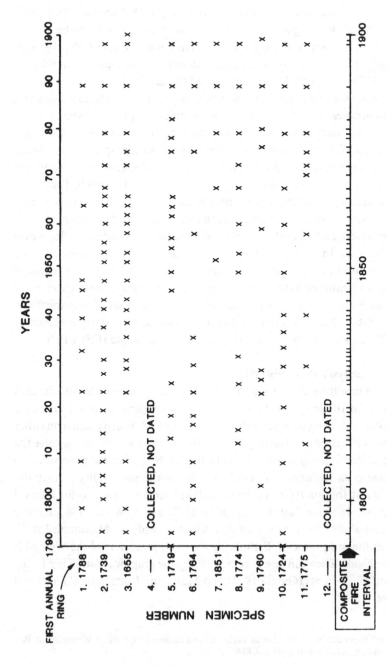

Figure 6.4. A fire event plot showing specimens and years in which fire evidence was found. The composite fire intervals which have the best evidence in a sample (cf. Dieterich 1980).

Table 6.1. *Time-since-fire distribution calculations from the stand origin map for the unlogged areas of the Boundary Waters Canoe Area, Minnesota (from Heinselman 1973)*

Age	Area in 1973 (ha)	Proportions of total area	Cumulative proportions
1–20	874	0.0052	0.99
21–40	3225	0.0192	0.98
41–60	1159	0.0069	0.97
61–80	54 947	0.3266	0.64
81–100	41 164	0.2446	0.40
101–120	39 485	0.2347	0.16
121–140	1075	0.0064	0.16
141–160	5607	0.0333	0.12
161–180	9343	0.0555	0.07
181–200	175	0.001	0.07
201–220	4973	0.0296	0.04
221–240	376	0.0022	0.03
241–260	1379	0.0082	0.03
261–280	97	0.0006	0.03
281–300	4060	0.0241	0.002
301–320	0	0	0.002
321–340	26	0.0002	0.002
341–360	0	0	—
361–380	298	0.0018	0.0
	168 262	1.00	

Fire history concepts

Besides the fire recurrence parameter (b) and the shape (variation) parameter (c) several other useful descriptions of the fire frequency distribution have come into use. Each class of concept comes as a pair, one the inverse of the other. Also, each concept can be expressed on a per sample basis or as a proportion of the whole universe studied (cf. Johnson and Van Wagner 1985).

The **Fire Cycle** (universe measure) is the time required to burn an area equal in size to the study area. Note that each sample need not burn once, just an *area* equal to the study area. The **Average Fire Interval** (element) is the expected return time per stand.

$$FI = FC = \int_0^\infty A(t)\,dt \qquad (6.6)$$

which is for the Weibull

$$bF(1/c+1)$$

where Γ is a gamma function. For the negative exponential, it is b.

The **Annual Percent Burn** is the proportion of the universe that burns per unit time and the **Fire Frequency** is the probability of an element burning per unit time:

$$\text{APB} = \text{FF} = \cfrac{1}{b\,\Gamma\!\left(\dfrac{1}{c}+1\right)} \qquad \text{Weibull}$$

(6.7)

$$\frac{1}{b} \qquad \text{negative exponential}$$

Annual percent burn can be used to convert the time-since-fire distribution into the Age Distribution of Stands Across the Landscape $A^*(t)$ (name due to C.G. Lorimer, pers. comm. 1985) as follows:

$$A^*(t) = \text{FF}\exp\left(-\left(\frac{t}{b}\right)^{c}\right) \qquad \text{Weibull}$$

(6.8)

$$\text{FF}\exp\left(\frac{1}{b}\right) \qquad \text{negative exponential}$$

The **Average Age of Stands Across the Landscape** is the centroid of the distribution and the **Average Prospective Life Time** of an element is the average time between fires. For the negative exponential it is b and

$$\text{AASA} = \text{APL} = \frac{\displaystyle\int_0^\infty tA(t)\,\mathrm{d}t}{\displaystyle\int_0^\infty A(t)\,\mathrm{d}t} = B\frac{\Gamma(2/c)}{\Gamma(1/c)} \qquad \text{Weibull} \qquad (6.9)$$

Mixed fire frequencies

Up to this point, we have assumed that within the region used in the fire history, there were no changes in fire frequency. This issue is of fundamental importance since fire ecology has postulated that topography (e.g. Heinselman 1973, Habeck 1976, Zackrisson 1977), fuel (e.g. Yarie 1981, Romme 1982, Romme and Knight 1981, DeSpain 1983), elevation and climate (Johnson 1979, Tande 1979, Arno 1980, Hemstrom and Franklin 1982) are important in defining the spatial or temporal boundaries of regions having similar fire frequencies. The fire history model just given assumes a region with a homogeneous fire frequency. Therefore, without a procedure for determining homogeneous distributions, fire cycle and other fire history concepts cannot be calculated.

Several methods exist for separating mixed fire frequencies. A graphic method is given by Kao (1959) and has been used by Johnson *et al.* (1990) and in modified form by Masters (1990). Mathematical techniques also exist (e.g. Falls 1970), but are much more difficult and often involve graphic estimation techniques in them.

A graphic procedure for recognizing and separating mixed fire frequencies into two homogeneous components is outlined below. (1) Plot the time-since-fire distribution on Weibull probability or semi-log paper in order to identify breaks in the slope which indicate changes in fire frequency (see Figure 6.5 for example of breaks); then (2) partition the mixed fire frequency using the graphic method of Kao (1959); (3) plot the individual components of the mixed fire frequency; and (4) correlate the component fire frequency to causes of the change in fire frequency. Step 1 has already been described. Step 2, the graphic method of partitioning, is as follows. Starting at each end of the graph, draw a line tangent to the dots. These two (or more) lines represent the estimate of the two (or more) distributions. By tracing from the intersection of the lines to the right, the percentage of samples in each distribution can be read. By multiplying this percentage by the total number of dots, the number of dots in each new distribution is determined. Step 3, the new distributions are then plotted as a cumulative percentage of *their* total. For step 4, there are no exact rules except that independent evidence is needed to test the hypothesized causes of the changes in fire frequency.

Boreal fire histories

Fire history data fall into three general categories: those with complete or statistically valid samples of the stand origin map, those that have fire records with fire maps (not stand origin maps) for recent years, and finally those that have small samples, no stand origin map, and unclear or incomplete sampling procedures, etc. The first of these categories allows construction of complete time-since-fire distributions, the second allows construction of the time-since-fire distribution only as far back as the fire maps allow (fire maps rarely go back beyond 30 years) and the final category does not allow a time-since-fire distribution to be constructed.

Four fire history studies in the boreal forest allow construction of time-since-fire distributions for at least 100 years (Figure 6.6). There are a large

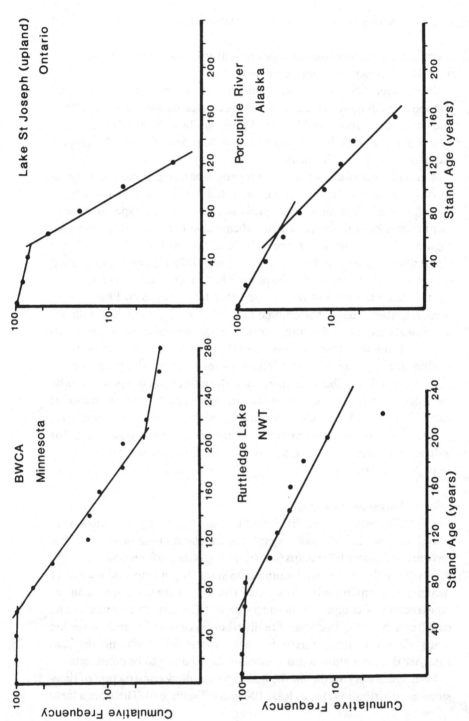

Figure 6.5. Mixed time-since-fire distributions for four fire history studies in the boreal forest plotted on semi-log paper. Boundary Waters Canoe Area data from Heinselman (1973), Ruttledge Lake from Johnson (1979), Lake St Joseph from Suffling *et al.* (1982) and Porcupine from Yarie (1981).

number of other studies which are incomplete in some manner and do not allow a time-since-fire distribution to be calculated. Some of these studies have calculated fire frequencies or fire cycles but without a fire frequency distribution it is difficult to interpret these values, even though they may be correct.

All four fire histories show a change in fire frequency 30–70 years ago and in two cases, Boundary Waters Canoe Area (Minnesota) (BWCA) and Ruttledge Lake (NWT), the record is long enough that a second fire frequency change 200 years ago is suggested. When the reason for the 30–70 year ago change was commented on in the individual studies, it was often given as better fire suppression activity. However, the closeness in timing of the changes and the low level of fire suppression in several of the areas (Ruttledge, Porcupine (Alaska), St Joseph (Ontario)) make this argument unconvincing. It is possible that the change in fire frequency detected early in this century is merely an artifact of some part of the fire history methodology. However, a more likely explanation is that the change is related to the long-term downward trend in fires noted in this century.

The fire records for all of Canada go back as far as 1918. Van Wagner (1988a) has assembled these data and made certain conservative additions for regions missing during the first three decades. Figure 6.6 shows that despite much variation from year to year the smoothed fire area values have a downward trend into the late 1970s. It is generally believed (Armstrong and Vines 1973, Harrington 1982, Van Wagner 1988a) that this downward trend is related to a climate in this century less conducive to fires, although an exact mechanism is not suggested. The increase of fires recorded in 1979–81 (Figures 6.6 and 3.1) exceeds any value in the 60 years of record and occurred despite better suppression capabilities. Although it is tempting to believe that this period represents the beginning of a change to more fires, it is more likely that this is the recurrence of a particularly bad set of years which are normal in any homogeneous fire frequency. Our problem is that since the fire cycle in the boreal forest is around 100 years, our present day records are not long enough.

The Boundary Waters Canoe Area and Ruttledge Lake fire history studies indicate a fire frequency change in the mid-1700s which may in some way be related to the Little Ice Age. A similar change in the mid-1700s has been identified in fire history studies in the Canadian Rockies (Johnson and Fryer 1987, Masters 1989), the Columbia Mountains of British Columbia (Johnson *et al.* 1990), and in northern Minnesota (Clark 1990). Payette *et al.* (1985) have also seen changes related to the Little Ice Age in dendroclimatic and fire history studies in northern Quebec.

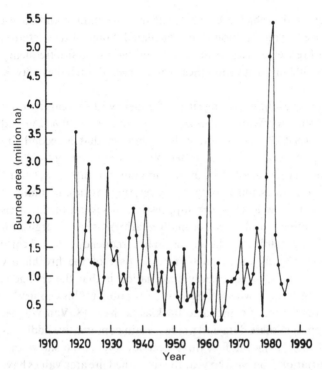

Figure 6.6. The area burned from 1918 to 1986 for all forested regions of
Canada, not just the boreal forest (cf. Van Wagner 1988a).

   The widespread occurrence of mixed fire frequency distributions in the
boreal forest, regardless of their causes, suggests that the mosaic of ages is
not in equilibrium with the present fire frequency. Large portions of the
present forest were established under a different fire frequency regime and
possibly different intensity and duff consumption regimes as well. When the
mixed fire frequency distributions in Figure 6.5 were partitioned into
homogeneous distributions, all fit negative exponentials and had fire cycles
from 40 to 60 years. The fire cycle, by definition, suggests that most (63%)
stands will never live much beyond the age at which stand canopy closure
occurs and very few will reach anything resembling old age.
   The negative exponential hypothesizes that the hazard of burning is
constant and that no change in flammability with age is occurring. The fact
that: (1) the areas over which these homogeneous fire frequencies occur are
large; (2) the fire cycles are short; and (3) there is a change in frequency in
the recent past, suggests that the factors which control the hazard of
burning are landscape level processes, probably climatic (see also Chapter 2).

Age mosaic patterns in the Stand Origin Map

In order for a more complete understanding of fire history, not only the survivorship (fire frequency), but also the spatial relationship between stand ages must be known. Since the fire frequency models suggest that the hazard of burning has a characteristic aging pattern (see equation 6.5), they also hypothesize a pattern in the adjacent stand ages in the stand origin map. Thus, the models hypothesize about how the burns create the age mosaic.

If the hazard of burning in the Weibull increases with age (shape parameter $c > 1$), then fires should start and/or spread in 'older' stands but tend not to burn into 'younger' stands. This would result in a pattern of adjacent old–young ages in the stand origin map. If the hazard is constant with age (Weibull shape parameter $c = 1$, i.e. a negative exponential), then fire start and/or spread should be indifferent to stand age and adjacent stand ages will show a random pattern. Finally, if the hazard of burning decreases with age (Weibull $c < 1$), then fires will start and/or spread in 'young' stands but tend not to burn into 'older' stands. A pattern of adjacent young–old stand ages should result, the same as when the hazard increases with age.

The best documented example of an ecosystem with an old–young mosaic resulting from an increasing hazard of burning with age is the chaparral of the USA and Mexico (Hanes 1971, Parsons 1976, Green 1981, Minnich 1983, Riggan *et al.* 1988). During the dry summer fire season, the age of chaparral stands plays a significant role in determining the hazard of burning. Live plants, even in the dry season, do not burn easily without substantial heat input which can be supplied only by dead stems, surface litter and dried grass and forbs. Chaparral shrubs younger than approximately 20 years since the last fire are still actively growing, and although mortality is occurring from thinning (Schlesinger and Gill 1978), there is still not a large accumulation of fine to medium drying fuels. In addition, the closed canopy tends to decrease both the ground cover plants and the accumulation of these fine fuels. After about 30 years, chaparral shrub growth decreases and mortality increases. Fine and medium drying fuels now accumulate as do fine ground fuels because of the opening of the canopy. Ignition and fire spread are now constrained not by lack of fine and medium drying fuels but by the weather. Consequently, fires tend to start and spread mostly in older stands. When younger stands are reached, the fire often extinguishes. Under natural conditions (no fire suppression), the

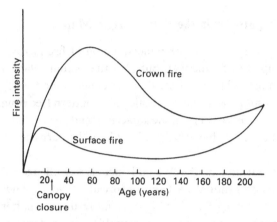

Figure 6.7. Suggested fire intensity pattern in a boreal forest stand as it ages (modified from Van Wagner 1983).

stand origin map has many small recent fires adjacent to younger stands which did not burn (Minnich 1983). Interestingly, the young–old pattern disappears in chaparral when fire suppression allows the entire landscape to become older than 20–30 years. Fires then become very large and are only stopped by factors unrelated to stand age (weather, topography, etc.).

The boreal forest seems to act like the chaparral system with fire suppression since its natural fire frequency is long enough to allow for adequate fine and medium fuel accumulation. In the few upland boreal forest stand origin maps tested for spatial autocorrelation with Merton's I (for equations, see Cliff and Ord 1981), none showed any young–old stand correlation. This is consistent with the constant hazard of burning ($c = 1$ negative exponential) in their time-since-fire distributions.

In the years immediately following a crown fire, spruce and pine stands accumulate a reasonable amount of fine fuels such as grasses and forbs (Figure 6.7). The ground may also be covered with needles that were killed by the fire but not consumed. Fires are possible in these early years, but as soon as a reasonable canopy of shrubs and saplings forms, the ground fine fuels are kept shaded and thus moist except in the driest of times. As the saplings grow in height, the ground and crown fire fuels separate and ground fires must now produce flames of increasing length (higher intensity) in order to ignite the crowns. Increasing the surface intensity (I_0) can be accomplished by increasing the available fine and medium drying fuels (particularly by longer periods of hot–dry weather) and/or by increasing the rate of spread. Consequently, the crown fire potential peaks soon after the ground fuel peaks. The longer the forest ages, the more the crown becomes elevated and the higher is the surface intensity required for crowning.

The fire frequencies in Figure 6.5 indicate that approximately 10% of the areas in the boreal forest are less than 20 years old (the average age at which canopy closure begins) and 75% of the areas are between 20 and 75 years. Thus, most of the boreal forest is in the period of high crowning potential. Rigorous studies of patterns of fires have been almost totally neglected and certainly provide an area of useful ecological research. However, to be useful the pattern studies must be firmly based on empirical understandings of fire behavior.

Fire frequency and population dynamics

If the fire frequency studies reported here are at all characteristic, from 0.5 to 2% of the upland boreal forest is burning per year. With this relatively short fire recurrence two questions arise: given both the fire frequency distribution and the plant survivorship curves (from causes other than fire), what is the chance of plants not dying from fire, and what is the chance of a plant surviving to reproduce before a fire? The first question tells us something about the relative importance of fire as a cause of mortality and the second question tells us if the population could reproduce itself.

To start, assume that the fire is lethal to the whole plant population, that both the fire frequency and plant survivorship from causes other than fire are at least approximately described by negative exponentials and that the two distributions are independent of each other. The probability of surviving without a fire is the probability that the plant mortality from causes other than fire (P) is not exceeded by the fire frequency (F) where

$$F_P(t) = \lambda_P \exp[-\lambda_P(t)]$$
$$F_F(t) = \lambda_F \exp[-\lambda_F(t)]$$

The chance of surviving the fire is:

$$R = \int_0^\infty f_F(t)\,[\int_F f_p(t)dt]dt$$
$$= \int_0^\infty \lambda_F \exp[-\lambda_F(t)][\exp(-\lambda_P(t)]dt$$
$$= \frac{\lambda_F}{\lambda_P + \lambda_F}$$

And if $\bar{P}=1/\lambda_p$ and $\bar{F}=1/\lambda_F$ then:

$$R=\frac{\bar{P}+\bar{F}}{\bar{P}} \qquad (6.10)$$

or in words, the probability of a plant surviving without a fire is the average mortality from causes other than fire divided by the sum of the average mortality from other causes and from fire.

From equation (6.10) it follows that when $\bar{P}=\bar{F}$, $R=1.0$, when $\bar{P}>\bar{F}$, $R>1.0$ and when $\bar{F}>\bar{P}$, $R<1.0$. Consequently, a plant population which has an average life span greater than the fire return interval has a better chance of surviving than a population with average life span less than the fire interval. Plants with life spans shorter than the fire interval will tend to become locally extinct before the fire occurs. Thus, to be represented in the post-fire forest, a species must either disperse into the site or be present as dormant buried seeds.

In southeastern Labrador, the cooler and wetter maritime climate has created a fire frequency much longer than the rest of the boreal forest. In these forests, Foster (1983, 1985; Foster and King 1986) has found widespread canopy tree senescence. The opening of the canopy allows the more shade tolerant understory balsam fir and black spruce to prosper. These forests act dynamically, as do other boreal forests in which the canopy is removed by logging, insects or disease, and the understory is not seriously disturbed. See Chapter 7 for further discussion of arboreal succession.

Further, what is the chance of a tree surviving a fire until it is old enough to reproduce? The probability of a fire during the period n is

$$P=1-(1-1/b)^n \qquad (6.11)$$

where b is the fire cycle. If n is the age of first reproduction, Figure 6.8 gives the probability of fire before the first reproductive age of the boreal forest trees. Jack pine, balsam fir, paper birch, and black and white spruce all have a good chance of being reproductive before a fire occurs.

Paleoecological studies of laminated lake sediments also give evidence of the effects of variation in fire intervals on regional vegetation. Swain (1973, 1980) in the Boundary Waters Canoe Area, Minnesota, and Larsen (1989) in Wood Buffalo National Park, Alberta, came to the same two general conclusions. Firstly, when fires occur at shorter intervals than the pollen producing age, species are reduced or absent in the ensuing vegetation. Secondly, longer-term changes in the fire frequency (perhaps as we have

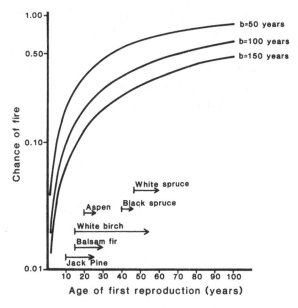

Figure 6.8. The probability of having no forest fire before a species' first reproduction. Age of first reproduction follows Fowells (1965).

seen in Figure 6.5) result in shifts in the predominant taxa over the period of the new frequency. Both authors have speculated that changes in fire intensity and duff removal were covariate with the change in frequency and might result in the changes in taxa abundance.

Although not part of the boreal forest, studies by Clark (1988, 1989, 1990) on fire frequency using charcoal buried in lake sediments deserve some discussion here. The use of charcoal fragments in varved lake sediments, pioneered by Swain (1973), has been significantly improved in recent years (e.g. Patterson *et al.* 1987). Particularly useful have been attempts to define closely the catchment from which charcoal fragments are being collected. Clark (1988, 1989, 1990), working in the red and white pine–northern hardwoods forest of the Lake Itasca region of northwestern Minnesota, USA, has used a charcoal stratigraphic analysis technique and a model of heavy particle dispersion which defined the charcoal catchment size to give a record of fire occurrence. By using the fire frequency models of the preceding sections, Clark found that fire frequency had changed from a wetter–cooler, less frequent fire period between AD 1240 and 1400, to a warmer–drier, more frequent fire period between 1400 and 1600 and a cooler–wetter period from 1600. He also noticed a period of decreased fire frequency since the early 1900s but attributed this to fire suppression. The similarity of Clark's changes in fire frequency to those reported in the preceding pages (Figure 6.5) again suggests that larger changes in fire frequency are controlled by large-scale atmospheric circulation patterns, while shorter-term changes in fire frequency are controlled by more local climate, landform and land-use patterns.

7

Fire and the population dynamics of boreal trees

Chapters 2 to 6 gave some of the physical reasons why the upland boreal forest is characterized by large, crown fires which remove large amounts of the forest floor organic matter and recur about every 100 years. In brief, the fire behavior results from the large conifer component of the forest which is an ideal fuel consisting of large numbers of small needles and branches, long canopy lengths, relatively short distances from the forest floor to the lower canopy, low foliage moisture, and a well-aerated forest litter and duff layer made up of small branches, decay resistant needles and cones. The airstream and synoptic weather patterns create the spring and summer temperature and precipitation patterns for fuel drying and lightning for ignition. Although this fire behavior is characteristic of large parts of the North American boreal forest, there are areas where it is not, particularly the maritime regions of Quebec, Labrador and Newfoundland.

This chapter will discuss how the fire behavior characteristics are reflected in tree recruitment and mortality patterns, and these result in a limited number of age distributions.

Some population background for boreal trees

Tree populations are never studied by following the individuals throughout their long lives. Instead, assumptions are made about the population structure and its birth, death and migration processes. These assumptions are validated, often with varying difficulty, and the age distribution then interpreted using these assumed population processes. See the Appendix at the end of this chapter on assumptions about the equivalence of age and survivorship distributions. For a general introduction to the ecology of boreal trees, see Ritchie (1987).

As a simple starting point, a population can be considered to be a

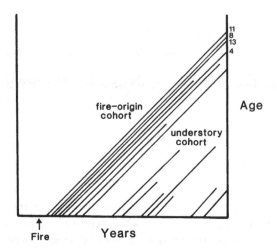

Figure 7.1. Lexis diagram showing the year and age of a population of jack pine. Each diagonal line represents one or more individuals (as marked) from its recruitment to the present (vertical line on right) or if it dies, to the point where the line stops. Two recruitment cohorts are recognized.

homogeneous group of individuals with respect to their birth and death schedules. If this assumption is unrealized, as it is in most boreal trees, then the population is divided into subpopulations called cohorts which are more homogeneous with respect to birth and death schedules. A birth or recruitment cohort is defined by the year(s) in which the subpopulation was added to the population. Birth or recruitment is an arbitrary term which can mean: seeds ripened on the parent tree; seeds actually dispersed to a spot on the ground from a specific or unspecified parent; or the germination of seeds from a specific or unspecified parent. Here we will use recruitment, not birth, to represent the rather vague idea of seed germination from undetermined parents. This leaves our population open to some immigration from trees a short distance away as well as the seeds from local trees.

The distribution of mortality is determined by recording the times of death of individuals in a cohort. The mortality rate is consequently the number dying during a time interval divided by the number alive at the start of the interval. A population is thus a collection of stacked recruitment cohorts. This diagonal stacking can be seen in a Lexis diagram (Figure 7.1).

In boreal trees, two general types of cohorts can be recognized: a **fire-origin cohort** which arises after a fire when canopy trees and ground cover plants have been killed and large amounts of duff have been ashed, and an **understory cohort** which arises after the fire cohort has formed a canopy and the forest floor is covered by ground cover plants. Each of these two cohort

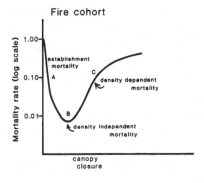

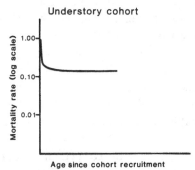

Figure 7.2. Above: mortality rate of fire cohort. A is the establishment mortality phase, B is the brief period when trees are still too small to crowd each other and suffer mortality by chance, C is the increased mortality rate due to thinning. The rate may decrease once trees have more evenly spaced themselves due to thinning mortality.
Below: mortality rate of an understory cohort showing the high mortality during establishment and then high and constant rate afterwards.

types generally has different recruitment and mortality schedules (Figure 7.2). Fire-origin cohorts for rapidly growing, shade intolerant trees such as jack pine usually have high recruitment and a U-shaped age-specific mortality rate. The U-shaped mortality rate results from high establishment mortality, followed by heavy thinning (crowding), then reduced thinning when mortality has more regularly spaced the trees, and finally increased rate from senescence–mortality. For slower growing, more shade tolerant species such as white spruce, the fire cohort may have a constant mortality as is typical of understory cohorts, but with a lower rate. Understory cohorts for both shade tolerant and intolerant species generally have very low recruitment and high, but relatively constant, age-specific mortality rates.

Recruitment requires seed-producing trees and a nearby seedbed suitable for germination. In boreal conifers, after reaching the minimum age of first reproduction, canopy position, not age, determines if an individual *is* reproductive. Trees in dominant canopy positions produce more seeds than do subcanopy and understory trees (e.g. Waldron 1965). Consequently, the fire-origin cohort has more reproductive members than understory cohorts because of their dominant canopy position.

Jack pine and black spruce have solved the dispersal problem into burns by having, respectively, serotinous and semi-serotinous cones. In jack pine, a resin bond holds the cones closed until the temperature is about 45 °C (Cameron 1953). Seeds remain viable in the cones for at least 10 years (Fowells 1965). In black spruce, cones at the end of branches are open, while cones near the main stem (i.e. on shorter plagiotropic branches) become covered with resin and open when the resin melts (e.g. Carleton 1982b). Serotinous and semi-serotinous cones provide a seed bank protected from the main heat of the fire and become available for dispersal after the fire. Seeds are released in pine and spruce cones for 2 or 3 years after a fire as the cones progressively reflex (Waldron 1965, Hellum and Pelchat 1979). These seeds remain viable in the soil for less than 3 years (Thomas and Wein 1985b).

White spruce, paper birch, aspen and balsam fir must disperse into a burn from surviving individuals or from adjacent unburned areas. White spruce and balsam fir have no alternative but to disperse while birch and aspen may survive fire and resprout from underground vegetative organs. White spruce and fir, because of the need for nearby seed trees, often may become locally extinct. Problems of dispersal were discussed in Chapter 3.

All boreal forest trees germinate best on exposed mineral soil (Vincent 1965, Fowells 1965, Cayford *et al.* 1967, Dobbs 1972, Cayford and McRae 1983), but have differing successes on duff. This differentiation of species in recruitment, depending on the amount of duff removed by fire, could be as important as canopy gap size is in deciduous forests. Tree-fall gaps in the deciduous forest are exploited by an array of trees which are adapted to various degrees of shade, canopy closure and surface disturbance (e.g. Hibbs 1982, Runkle 1985). The boreal forest trees may also show an array of tree adaptations for recruitment or vegetative regrowth on different amounts of duff and shade from recovering and surviving herbs and shrubs (Thomas and Wein 1985a). For example, jack pine has rapid seedling and sapling growth rates and little shade tolerance. Its establishment rate decreases rapidly on surfaces with any duff remaining (Chrosciewicz 1974). Black spruce (on uplands) has a relatively slow seedling and sapling growth rate, but is moderately shade tolerant. It establishes better when some duff remains and when surviving or regenerating shrubs provide shade. Balsam fir has the slowest seedling and sapling growth rates, is the most shade tolerant and establishes better on duff and in shrub and herb shade than any of the other trees. White spruce seems to be intermediate between black spruce and balsam fir. Notice that the changing species preferences along a gradient of decreasing duff consumption and increasing shrub and herb cover also reflect the gradient of site moistures that the species prefer (Vincent 1965, Carleton and Maycock 1978, 1980, 1981, Black and Bliss 1980, Carleton 1982a).

Age distribution of boreal trees

In the upland boreal forest, a limited number of species age distributions are realized (see Whipple and Dix 1979 for a different vegetation type). Age distributions are used here instead of diameter distributions since there is often a poor correlation between age and diameter, particularly for canopy subdominant and understory trees (Lorimer 1985). Further, the age distributions used are from single species and stands rather than sets of stands from disparate locations and/or more than one species added together, since the local single species population is the object of interest, not a regional mixed population (see also Stephenson 1987). Age distributions can be divided into two groups for the purposes of discussion: single and multiple cohort distributions.

Single cohort age distributions

A single cohort age distribution (Figure 7.3*A*) consists of one group of individuals which recruited for a short period after a fire. Jack pine and black spruce sometimes have this kind of age distribution (e.g. Yarranton and Yarranton 1975, Carleton and Wannamaker 1987). This narrow recruitment window results from the co-occurrence of seed supply, canopy removal, duff removal and suitable establishment conditions. Recruitment stops when the seedbed is again covered by shrubs, herbs, mosses and litter. In actual fact, this single cohort distribution may be rare and the two cohort distributions discussed later may be more characteristic. The interpretation of two cohort distributions as a single cohort distribution results from not sampling seedling and small saplings.

After recruitment, age-specific mortality follows a U-shaped distribution. If densities are high, competition (thinning) will start after 15–30 years

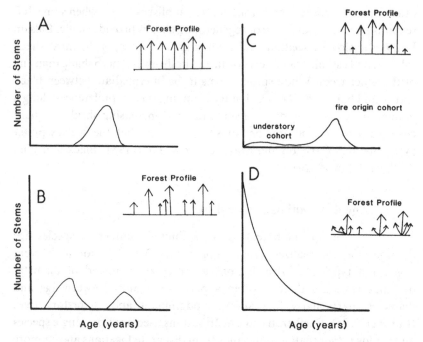

Figure 7.3. *A*. Single cohort age distributions which arose after a fire. *B*. A series of separate cohorts which arose after low intensity fires killed only some of the canopy trees and exposed mineral soil so a cohort of trees could become established. Usually only bimodal age distributions are observed. *C*. An age distribution which is a mixture of two or more cohorts. *D*. All aged distributions in open stands with vegetative reproduction.

(Evert 1970, Yarranton and Yarranton 1975, Carleton and Wannamaker 1987, and others) and often a bimodal distribution of diameters will develop in the cohort (Mohler *et al.* 1978). This diameter stratification in the cohort gives the impression that the smaller diameter trees are younger than the larger diameter trees. Also, when the diameter distributions of different species are compared, a similar confusion can arise. For example, young jack pine has more rapid height growth than black spruce of the same age. Consequently jack pine, being larger, is viewed as the pioneer tree while black spruce is seen as recruiting later in the shade of the pine. However, on some sites the black spruce may eventually overtop the jack pine. Oliver (1981) has called attention to this behavior in many conifer and deciduous trees.

Multiple cohort age distribution

The simplest multiple cohort age distribution is that with several clearly separated cohorts (Figure 7.3*B*). These separate cohorts arise after passive crown fires cause sufficient canopy mortality and duff removal to allow brief periods of recruitment (Carleton 1982b). These stands are recognized by the fire killed (as opposed to thinning killed) individuals on the forest floor and fire scars on older individuals associated with the dates of recruitment. Age-specific mortality for each cohort usually follows a U-shaped curve.

The next multiple cohort age distribution (Figure 7.3*C*) is very similar to the single cohort distribution. This is best called a two cohort distribution because it contains a fire-origin and understory cohort (Carleton 1982b, Johnson and Fryer 1989, Sirois and Payette 1989). The fire-origin cohort follows a mortality pattern like the single cohort age distribution. The understory cohort(s) has very few members and very high mortality. Some may persist in suppressed growth states for long periods (particularly, more shade tolerant species such as black spruce and balsam fir). There is a very low probability that members of the understory cohort will survive to replace the fire cohort canopy. This is because of their low numbers, high mortality, inability at older ages to change to a rapid height growth when a canopy opening occurs and the short interval (approximately 100 years) between fires. Unlike deciduous forests where understory-origin cohorts form advanced regeneration if a canopy opening occurs, most boreal forest openings occur from fires which kill the understory and canopy.

A clearly documented exception to this is regeneration of balsam fir following spruce budworm epidemics (Blais 1983). In this situation the canopy balsam fir and/or spruce is progressively killed over several years by spruce budworm (*Choristoneura fumiferana* (Clem)) outbreaks and by blowdown of the dead or weakened spruce budworm infected trees. The opening of the canopy allows the vigorous growth of understory herbs, shrubs, and already established understory trees. This advanced tree regeneration then grows up to create a new canopy (Baskerville 1975, Morin and Laprise 1990, Morin 1990). Large outbreaks of spruce budworm are common in regions to the south of the boreal forest proper but are less common as one moves north (Blais 1983, Hardy *et al.* 1983, 1985), probably because of cooler temperatures and slower development of the insect (Hardy *et al.* 1983). In the boreal forest, spruce budworm outbreaks are more limited, tending to produce patches of infected trees which then are subject to blowdown (Morin 1990). The relationship of fire and spruce budworm outbreaks is still not well understood. It is generally believed that budworm caused patches of mortality, blowdown and regeneration occur within the normal fire cycle for the region. Limited evidence suggests that budworm killed stands have a higher fire hazard for only a few years before the herbaceous understory growth decreases the hazard (Stocks 1987b).

The final multiple cohort distribution considered is the all-aged distribution (Figure 7.3D). This distribution is often considered to indicate a self-reproducing population. In the boreal forest, such a distribution is found in open canopied stands, particularly in the lichen woodlands of the subarctic (Payette and Gagnon 1979, Légère and Payette 1981, Payette and Filion 1985, Payette *et al.* 1985). This age distribution arises from a combination of vegetative growth around single isolated individuals of black spruce or paper birch and infrequent episodes of recruitment following ground or passive crown fires which did not kill all of the older trees.

Delayed population recruitment

It is easy to see that there is a basic similarity between the different types of age distributions recognized in the last section. Boreal tree populations seem to have essentially one cohort which establishes after a fire and whose fitness is not realized until after the next fire. This does not mean that no recruitment occurs during the life of the fire-origin cohort, but that the recruitment will not survive the next fire to reproduce. Many of the age distributions have more than one cohort, but these understory cohorts have few members and little chance of reaching the canopy (and becoming reproductive) before the next fire.

The delay in offspring recruitment of a particular species' fire-origin cohort is determined by the fire behavior. The boreal forest fire frequency is short (average frequency is *c.* 0.01) so there is little chance that the fire-origin cohort will become locally extinct before a fire occurs. The fire intensity and duff removed will set limits on the amount of recruitment by determining canopy mortality, dispersal distance, and seedbed conditions. Of course the weather, seed predation, and herbivory, etc. after germination will determine seedling survival.

The delay in recruitment leads to changes in the stability of (local) populations from a fixed point to periodic and to stochastic variation in numbers (May 1973). Little empirical information exists for boreal tree populations over several fire events. Consequently, little can be said at this time about the nature of this variation except that fixed point and periodic solutions are probably not common. The stochastic nature of fire behavior parameters will probably lead to stochastic variation in populations. Paleoecological studies such as those discussed in Chapter 6 are of limited usefulness since they sample regional, not local, populations. However, studies of sites which sample the local pollen (Bradshaw 1988) would seem to offer a potential source of empirical data on long-term population dynamics.

Arboreal succession

Arboreal succession, the replacement of canopy trees by under-story of the same or different species, is not the usual occurrence in the boreal forest (Rowe 1961, Dix and Swan 1971, Carleton and Maycock 1978, Johnson 1981, Cogbill 1985, Bergeron and Dubuc 1989, and many others). Fire behavior instead separates one generation of trees from the next. Arboreal vegetation dynamics is thus reduced to understanding the population dynamics between fires with the fire behavior acting as the filter between one generation of trees and the next.

As we gain more information on the process–response patterns of fire behavior, we will probably find that the persistence or change in tree species on a site contains the uniqueness of an individual fire's behavior and the pre-fire vegetation. The landscape, on the other hand, will contain not only the variation in fire behavior over decades and kilometers, but also the tree responses to fire hundreds of years before. This may well create problems in the use of old stands as parts of present day chronosequences since these old stands may have resulted from different fire behavior patterns.

Between fires, the changes in tree canopy species which involve real difference in age (i.e. cohort structure) and not just height growth rates are limited (see references above). The understory seedlings and saplings, as we have seen, are very few in number and suffer high mortality. The chances of understory trees replacing the canopy are severely limited by the short fire recurrence time, the high fire intensity and large duff consumed. Hence, understory seedlings and saplings do not generally represent the next canopy generation (e.g. Carleton and Maycock 1978, Bergeron and Dubuc 1989), but instead represent trees with life history traits (e.g. shade tolerance, ability to germinate and establish on moss and herbaceous covered substrate) which allow recruitment and temporary survival. Balsam fir (Bakuzis and Hansen 1965) and to a lesser degree white spruce (Rowe 1970) are examples of trees with these life history traits. It should, however, be stressed that both fir and white spruce are also present in the fire cohort when a seed source is nearby and site conditions (both soil-topography and post-fire) are suitable.

Occasionally understory saplings may be able to act as advanced regeneration of the canopy when logging, insects (e.g. spruce budworm: Baskerville 1975, Blais 1983, Morin 1990, Morin and Laprise 1990) or old age (long intervals between fires) open the canopy (Foster 1983, 1985, Foster and King 1986).

Appendix: Age and survivorship distributions

Many tree population studies imply that the age distribution in some way reflects the survivorship distribution. The argument that an age distribution is equivalent to a survivorship distribution is as follows.

The age distribution $a(x)$ of a density independent population can be written as:

$$a(x) = \frac{B(t-x)\,lx}{\sum\limits_{x=0} B(t-x)\,lx} \qquad (A7.1)$$

where $B(t-x)$ is the number of plants of age class x at its recruitment t years ago and lx is the proportion surviving to age x.

If the age distribution $a(x)$ is applied to a population with only one cohort or with several cohorts all having the same age-specific recruitment and mortality rates, the age distribution will, after a short time, approach a constant proportion of individuals in each age class and the number in the whole population (i.e. all age classes) will increase, decrease or remain constant at a characteristic rate. This follows from equation (A7.1) if one remembers that a population with all cohorts following the same age-specific recruitment and mortality will grow at an exponential rate. Recruitment will then be

$$B(t) = \exp(rt)\,B_0 \qquad (A7.2)$$

where r is the instantaneous growth rate of density independent growth.

Now, by substituting (A7.2) into equation (A7.1):

$$a(x) = \frac{\exp(-rx)\,lx}{\sum\limits_{x=0} \exp(-rx)\,lx} \qquad (A7.3)$$

Notice that the value of r determines the shape of $a(x)$. If $r = 0$, then $a(x) = lx$ and the age distribution is the same as the survivorship. However, remember that this is only true when all cohorts have the same vital rates and the population size is constant. Unfortunately, multiple cohort populations in boreal trees rarely have either similar vital rates or constant population size.

8

Conclusion

It would seem that boreal forest fires take place over both a large landscape scale and long time span. As ecologists, we have found these spatial and temporal scales difficult to study rigorously. We are still unsophisticated in our understanding of fire behavior and its population effects in the boreal forest. Many of our most cherished ecological ideas of fire effects are not well documented and many are simply intuitive arguments with informal observational support.

At best, we can state that on average boreal wildfires are large, infrequent, of high intensity and consume large amounts of the forest floor duff. Despite substantial evidence that the physical characteristics and moisture of wildland fuels account for more variation in fire behavior than do variations in fuel chemistry, there is a persistent desire to state that flammability (a vague term: see Hilado 1977, Mak 1988) in natural vegetation is determined by natural selection for fuel chemistry. However, boreal forest fire behavior results primarily from: a vegetation which produces a large amount of relatively fine fuels ($\leqslant 2$ cm in diameter) which decays slowly; and fire seasons which, over the life span of a tree (on average c. 50–150 years), can be expected to have at least one synoptic weather pattern which severely dries the ground fine and duff fuels, has lightning which ignites the fuel and high winds which cause high rate of fire spread and intensity. Further, the crown architecture of conifers and their low foliar moisture make crown fires possible.

This brings us to the issue of ground–crown fire regimes. In most crown fire susceptible regions there is an alternation between small, infrequent, low intensity ground fires which consume ground fuels, but do not generally kill larger trees, and infrequent, large, high intensity crown fires which kill most of the canopy trees. In the boreal forest, most fire seasons are not very conducive to high intensity and rate of spread fires. Therefore, in most years fires are few, small and of low intensity. However, at infrequent intervals (on average every 100 years), weather conditions lead to high rates of spread and intensity fires. These fires are little affected by the few, small, low intensity burned areas they encounter. Thus, the vegetated landscape is predominantly the result of these infrequent, large, high intensity fires. The ground fire regime appears to cover such a small area that its influence on the regional vegetation should be limited. However, research on the local variation in duff consumption and surviving single or small groups of trees within large, high intensity fires has been almost completely neglected.

The negative exponential fire frequency and spatial autocorrelation studies support the idea that the mosaic of the time since fire (stand origin maps) has a chance pattern, i.e. the ages of contiguous mosaics are decided by chance. However, little is still known about the effects on the scale of the age patches in landscapes of different terrain (e.g. more lakes and bogs). Large numbers of studies have speculated on these effects (Heinselman 1973, Johnson 1979, Foster 1983). Foster (1983) has suggested that the low fire frequency and higher rainfall in southeastern Labrador have led to extensive peatland development on the uplands.

The intuitive idea that in the upland boreal forest small landscape areas, in the order of hectares, could *consistently* survive fires longer (shorter) within a matrix of *consistently* higher (lower) fires does not seem to have any rigorous proof at present. Certainly, small areas survive individual fires, but they do not seem to have a time-since-fire distribution (survivorship) which is significantly different from the matrix which surrounds them. One cannot arbitrarily separate out parts of a landscape, e.g. all white spruce stands, and calculate their fire frequency without showing that these stands' fire frequencies are significantly different and that they are subject to a different fire hazard, i.e. individual fires have time after time *consistently* different fire behavior in these stands. One way of determining this is to study the behavior and effects of individual fires (Fryer and Johnson 1988). Up to now, there are few fire effect studies which can bear on this question (e.g. Weber *et al.* 1987).

The large landscape areas which have homogeneous fire frequencies suggest that regional variables are responsible. Airstream and synoptic meteorological variables would seem to be the best candidates. The resulting high intensities and rates of spread create fire behavior which overrides most small landscape differences.

Unburned or less severely burned areas always occur within a fire. Eberhart and Woodard (1987) describe the pattern of these patches in one of the few studies in the boreal forest. In 69 fires in northern Alberta, they found that larger fires (> 2000 ha) were more elliptical in shape, suggesting they are wind driven (Anderson 1983), and the number of unburned patches increased in density and size when compared with smaller fires. These results suggest that unburned patches may be not only a function of natural fire barriers, but also related to the mechanics of fire spread itself (Ohtsuki and Keyes 1986, Neissen and Blumen 1988).

Finally, boreal fires recur at the time scale of the life span of the trees, kill most of the canopy and understory and create a good seedbed for boreal tree species. Population   should track the fire behavior (Hutchinson 1953, Caswell 1978, and many others). When fires occur, the populations are reduced and then have the possibility of rapid increases (note here we are speaking about population increase, not individual growth), usually by recruitment. Thus, fire behavior and population recruitment may control population abundance and distributions. To argue that by stopping fires a shade tolerant, self-reproducing tree composition would eventually result is to disregard the life histories of most boreal trees and, further, require a major change in climate and in vegetation architecture.

References

Ahlgren, C.E. (1960). Some effects of fire on reproduction and growth of vegetation in northeastern Minnesota. *Ecology* 41: 431–45.

Ahlgren, I.F. & Ahlgren, C.E. (1960). Ecological effects of forest fires. *Botanical Reviews* 26: 483–533.

Albini, F.A. (1975). An attempt (and failure) to correlate duff removal and slash fire heat. *US Department of Agriculture, Forest Service, General Technical Report INT-24.*

Albini, F.A. (1976). Estimating wildfire behavior and effects. *US Department of Agriculture, Forest Service, General Technical Report INT-30.*

Albini, F.A. (1981). A model for the wind-blown flame from a line fire. *Combustion and Flame* 43: 155–74.

Albini, F.A. (1984). Wildland fires. *American Scientist* 72: 590–7.

Albini, F.A. (1985). Wildland fire spread by radiation – a model including fuel cooling by natural convection. *Combustion Science and Technology* 45: 101–13.

Albini, F.A. (1986). Predicted and observed rates of spread of crown fires in immature jack pine. *Combustion Science and Technology* 48: 65–76.

Albini, F.A. & Stocks, B.J. (1986). Predicted and observed rates of spread of crown fires in immature jack pine. *Combustion Science and Technology* 48: 65–76.

Alexander, M.E. (1982). Calculating and interpreting forest fire intensities. *Canadian Journal of Botany* 60: 349–57.

Alexander, M.E. (1985). Estimating the length-to-breadth ratio of elliptical forest fire patterns. In *Proceedings of the Eighth Conference on Fire and Forest Meteorology, April 29–May 2, Detroit, Michigan*, pp. 287–304. Bethesda, Maryland: Society of American Foresters.

Alexander, M.E. (1988). Help with making crown fire hazard assessments. Proceedings of the Symposium and Workshop on Protecting People and Homes from Wildfire in the Interior West, pp. 147–56. *US Department of Agriculture, Forest Service, General Technical Report INT – 251.*

Alexander, M.E., Janz, B. & Quintillio, D. (1983). Analysis of extreme wildfire behavior in east-central Alberta: a case study. In *Preprint Volume of the Seventh Conference on Fire and Forest Meteorology, April 25–28, Fort Collins, Colorado*, pp. 38–46. Boston: American Meteorological Society.

Alexander, M.E., Lawson, B.D., Stocks, B.J. & Van Wagner, C.E. (1984). User guide to the Canadian forest fire behavior prediction system: rate of spread relationships. *Environment Canada, Canadian Forestry Service Fire Danger Group, Interim edition.*

Anderson, H.E. (1966). Mechanisms of fire spread. *US Department of Agriculture, Forest Service, Research Paper INT-28.*

112   *References*

Anderson, H.E. (1983). Predicting wind-driven wildland fire size and shape. *US Department of Agriculture, Forest Service, Research Paper INT-305.*

Archibold, O.W. (1979). Buried viable propagules as a factor in postfire regeneration in northern Saskatchewan. *Canadian Journal of Botany* **57**: 54–8.

Archibold, O.W. (1980). Seed input into a postfire forest site in northern Saskatchewan. *Canadian Journal of Forest Research* **10**: 129–34.

Archibold, O.W. (1984). A comparison of seed reserves in Arctic, Subarctic and Alpine soils. *Canadian Field Naturalist* **98**: 337–44.

Armstrong, J. & Vines, R.G. (1973). Possible periodicities in weather patterns and Canadian forest fire seasons. *Environment Canada, Forestry Service, Forest Fire Research Institute, Information Report FF-X-39.*

Arno, S.F. (1980). Forest fire history in the Northern Rockies. *Journal of Forestry* **78**: 460–5.

Arno, S.F. & Sneck, K.M. (1977). A method for determining fire history in coniferous forests of the mountain west. *US Department of Agriculture, Forest Service, General Technical Report INT-42.*

Artley, D.K., Shearer, R.C. & Steele, R.W. (1978). Effects of burning moist fuels on seedbed preparations in cutover western larch forests. *US Department of Agriculture, Forest Service, Research Paper INT-211.*

Bakuzis, E.V. & Hansen, H.L. (1965). *Balsam Fir, Abies balsamea (Linnaeus) Miller.* Minneapolis: University of Minnesota Press.

Barney, R.J. (1969). *Interior Alaska wildfires 1956–1965.* US Department of Agriculture, Forest Service, Pacific Northwest Forest and Range Experiment Station, Institute of Northern Forestry, Juneau, Alaska.

Barney, R.J. (1971). Selected 1966–69 interior Alaska wildfire statistics with long-term comparisons. *US Department of Agriculture, Forest Service, Research Note PNW-154.*

Baskerville, G.L. (1975). Spruce budworm: super silviculturist. *Forestry Chronicle* **51**: 138–40.

Bergeron, Y. & Brisson, J. (1990). Fire regime in red pine stands at the northern limit of the species' range. *Ecology* **71**: 1352–64.

Bergeron, Y. & Dubuc, M. (1989). Succession in the southern part of the Canadian boreal forest. *Vegetatio* **79**: 51–63.

Berlad, A.L. (1970). Fire spread in solid fuel arrays. *Combustion and Flame* **14**: 123–36.

Bishop, C.A. (1974). *The Northern Ojibwa and the Fur Trade: an Historical and Ecological Study.* Toronto: Holt, Rinehart & Winston of Canada, Ltd.

Black, R.A. & Bliss, L.C. (1980). Reproductive ecology of *Picea mariana* (Mill.) Bsp., at tree line near Inuvik, Northwest Territories, Canada. *Ecological Monographs* **50**: 331–54.

Blackwell, P.G., Rennolls, K. & Coutts, M.P. (1990). A root anchorage model for shallowly rooted Sitka spruce. *Forestry* **63**: 73–91.

Blais, J.R. (1983). Trends in the frequency, extent, and severity of spruce budworm outbreaks in eastern Canada. *Canadian Journal of Forest Research* **13**: 539–47.

Bliss, L.C. (1962). Caloric and lipid content in alpine tundra plants. *Ecology* **43**: 753–7.

Bradshaw, R.H.W. (1988). Spatially-precise studies of forest dynamics. In *Vegetation History*, ed. B. Huntley & T. Webb III, pp. 725–51. Belgium: Kluwer Academic Publishers.

Bray, J.R. & Gorham, E. (1964). Litter production in forests of the world. *Advances in Ecological Research* **2**: 101–57.

Brotak, E.A. & Reifsnyder, W.E. (1976). Synoptic study of the meteorological conditions associated with extreme wildland fire behavior. In *Proceedings of the Fourth National*

Conference on Fire and Forest Meteorology, November 16–18, St Louis, Missouri, pp. 66–9. US Department of Agriculture, Forest Service, General Technical Report RM-32.

Brown, J.K., Marsden, M.A., Ryan, K.C. & Reinhardt, E.D. (1985). Predicting duff and woody fuel consumed by prescribed fire in the northern Rocky Mountains. *US Department of Agriculture, Forest Service, Research Paper INT-337.*

Bryson, R.A. (1966). Air masses, streamlines, and the boreal forest. *Geographical Bulletin* **8**: 228–69.

Bryson, R.A. & Hare, F.K. (1974). *World Survey of Climatology. Volume 11: Climates of North America.* Amsterdam: Elsevier.

Byram, G.M. (1954). Atmospheric conditions related to blowup fires. *Station Paper No. 35. Southeastern Forest Experiment Station.* Asheville, North Carolina: US Department of Agriculture, Forest Service.

Byram, G.M. (1959). Combustion of forest fuels. In *Forest Fire: Control and Use*, ed. K.P. Davis, pp. 61–89. New York: McGraw-Hill.

Byram, G.M. (1966). Scaling laws for modeling mass fires. *Pyrodynamics* **4**: 271–84.

Byram, G.M. & Martin, R.E. (1970). The modeling of fire whirlwinds. *Forest Science* **16**: 386–99.

Calmes, M.A. & Zasada, J.C. (1982). Some reproductive traits of four shrub species in the black spruce forest type of Alaska. *Canadian Field Naturalist* **96**: 35–40.

Cameron, H. (1953). Melting point of the bonding material in lodgepole pine and jack pine cones. *Canada Department of Resources and Development, Forestry Branch, Silviculture Leaflet No. 86.*

Carleton, T.J. (1982a). The composition, diversity, and heterogeneity of some jack pine (*Pinus banksiana*) stands in northeastern Ontario. *Canadian Journal of Botany* **60**: 2629–36.

Carleton, T.J. (1982b). The pattern of invasion and establishment of *Picea mariana* (Mill.) BSP. into the subcanopy layers of *Pinus banksiana* Lamb. dominated stands. *Canadian Journal of Forest Research* **12**: 973–84.

Carleton, T.J. & Maycock, P.F. (1978). Dynamics of the boreal forest south of James Bay. *Canadian Journal of Botany* **56**: 1157–73.

Carleton, T.J. & Maycock, P.F. (1980). Vegetation of the boreal forests south of James Bay: non-centered component analysis of the vascular flora. *Ecology* **61**: 1199–212.

Carleton, T.J. & Maycock, P.F. (1981). Understorey–canopy affinities in boreal forest vegetation. *Canadian Journal of Botany* **59**: 1709–16.

Carleton, T.J. & Wannamaker, B.A. (1987). Mortality and self-thinning in postfire black spruce. *Annals of Botany* **59**: 621–8.

Caswell, H. (1978). Predator-mediated coexistence: a nonequilibrium model. *American Naturalist* **112**: 127–54.

Cayford, J.H., Chrosciewicz, Z. & Sims, H.P. (1967). A review of silvicultural research in jack pine. *Canadian Department of Forestry and Rural Development, Forestry Branch Departmental Publication No. 1173.*

Cayford, J.H. & McRae, D.J. (1983). The ecological role of fire in jack pine forests. In *The Role of Fire in Northern Circumpolar Ecosystems*, ed. R.W. Wein & D.A. MacLean, pp. 183–99. New York: John Wiley.

Chapman, L.J. & Thomas, M.K. (1968). The climate of northern Ontario. *Climatological Studies No. 6, Canada, Department of Transport, Meteorological Branch.* Toronto, Ontario.

Christensen, N.L. (1985). Shrubland fire regimes and their evolutionary consequences. In *The Ecology of Natural Disturbance and Patch Dynamics*, ed. S.T.A. Pickett & P.S.

White, pp. 85–100. New York: Academic Press.

Chrosciewicz, Z. (1974). Evaluation of fire-produced seedbeds for jack pine regeneration in central Ontario. *Canadian Journal of Forest Research* **4**: 455–7.

Chrosciewicz, Z. (1976). Burning for black spruce regeneration on a lowland cutover site in southeastern Manitoba. *Canadian Journal of Forest Research* **6**: 179–86.

Chrosciewicz, Z. (1986). Foliar heat content variations in four coniferous tree species of central Alberta. *Canadian Journal of Forest Research* **16**: 152–7.

Clark, J.S. (1988). Particle motion and the theory of charcoal analysis: source area, transport, deposition and sampling. *Quaternary Research* **30**: 81–91.

Clark, J.S. (1989). Ecological disturbance as a renewal process: theory and application to fire history. *Oikos* **56**: 17–30.

Clark, J.S. (1990). Fire and climate change during the last 750 yr in northwestern Minnesota. *Ecological Monographs* **60**: 135–59.

Cliff, A.D. & Ord, J.K. (1981). *Spatial Processes: Models and Applications.* London: Pion Ltd.

Cogbill, C.V. (1985). Dynamics of the boreal forests of the Laurentian Highlands, Canada. *Canadian Journal of Forest Research* **15**: 252–61.

Countryman, C.M. (1964). Mass fires and fire behavior. *US Department of Agriculture, Forest Service Research Paper PSW-19.*

Cox, D.R. (1962). *Renewal Theory.* London: Methuen.

Deans, J.D. & Ford, E.D. (1983). Modelling root structure and stability. *Plant and Soil* **71**: 189–95.

Deeming, J.E., Lancaster, J.W., Fosberg, M.A., Furman, R.W. & Schroeder, M.J. (1974). The national fire-danger rating system. *US Department of Agriculture, Forest Service, Research paper RM-84 (Revised).*

De Groot, W.J. & Alexander, M.E. (1986). Wildfire behavior on the Canadian shield: a case study of the 1980 Chachukew fire, east-central Saskatchewan. In *Proceedings, Third Central Region Fire Weather Committee Science and Technology Seminar,* ed. M.E. Alexander, pp. 23–45. Canadian Forestry Service, Study NOR-5-05.

Denny, M.W. (1988). *Biology and the Mechanics of the Wave-Swept Environment.* Princeton, NJ: Princeton University Press.

Despain, D.G. (1983). Nonpyrogenous climax lodgepole pine communities in Yellowstone National Park. *Ecology* **64**: 231–4.

Dieterich, J.H. (1980). The composite fire interval – a tool for more accurate interpretation of fire history. In *Proceedings of the Fire History Workshop,* Oct. 20–24, 1980, ed. M.A. Stokes and J.H. Dieterich. *US Department of Agriculture, Forest Service, General Technical Report RM-81.*

Diotte, M. & Bergeron, Y. (1989). Fire and the distribution of *Juniperus communis* L. in the boreal forest of Quebec, Canada. *Journal of Biogeography* **16**: 91–6.

Dix, R.L. & Swan, J.M.A. (1971). The roles of disturbance and succession in upland forest at Candle Lake, Saskatchewan. *Canadian Journal of Botany* **49**: 657–76.

Dobbs, R.C. (1972). Regeneration of white and Engelmann spruce: a literature review with special reference to the British Columbia interior. Canadian Department of the Environment, *Canadian Forestry Service, Pacific Forest Research Centre, Information Report, BC-X-69.*

Eberhart, K.E. & Woodard, P.M. (1987). Distribution of residual vegetation associated with large fires in Alberta. *Canadian Journal of Forest Research* **17**: 1207–12.

Evert, F. (1970). Black spruce growth and yield at various densities in the Ontario Clay Belt. *Forest Science* **16**: 183–95.

Falls, L.W. (1970). Estimation of parameters in compound Weibull distributions.

Technometrics **12**: 399–407.

Fang, J.B. (1969). An investigation of the effect of controlled wind on the rate of firespread. PhD dissertation, University of New Brunswick, Fredericton, Canada.

Ferguson, E.R. (1955). Fire-scorched trees: will they live or die? In *Proceedings of the Fourth Annual Forestry Symposium*. School of Forestry, Louisiana State University, Baton Rouge, Louisiana.

Flannigan, M.D. & Harrington, J.B. (1988). A study of the relation of meteorological variables to monthly provincial area burned by wildfire in Canada (1953–80). *Journal of Applied Meteorology* **27**: 441–52.

Flinn, M.A. & Wein, R.W. (1977). Depth of underground plant organs and theoretical survival during fire. *Canadian Journal of Botany* **55**: 2550–4.

Fons, W.L. (1946). Analysis of fire spread in light forest fuels. *Journal of Agricultural Research* **72**: 93–121.

Fosberg, M.A. (1975). Heat and water vapor flux in conifer forest litter and duff: a theoretical model. *US Department of Agriculture, Forest Service, Research Paper RM-152*.

Foster, D.R. (1983). The history and pattern of fire in the boreal forest of southeastern Labrador. *Canadian Journal of Botany* **61**: 2459–71.

Foster, D.R. (1985). Vegetation development following fire in *Picea mariana* (black spruce) – *Pleurozium* forests of south-eastern Labrador, Canada. *Journal of Ecology* **73**: 517–34.

Foster, D.R. & King, G.A. (1986). Vegetation pattern and diversity in S.E. Labrador, Canada: *Betula papyrifera* (Birch) forest development in relation to fire history and physiography. *Journal of Ecology* **74**: 465–83.

Fowells, H.A. (1965). Silvics of forest trees of the United States. *US Department of Agriculture, Forest Service, Agriculture Handbook 271*.

Fritts, H.C. (1976). *Tree Rings and Climate*. London: Academic Press.

Franklin, J.F. & Smith, C.E. (1973). Seeding habits of upper-slope tree species. II. Dispersal of mountain hemlock seedcrop on a clear-cut. *US Department of Agriculture, Forest Service, Research Note PNW-214*.

Franklin, J. F. & Smith, C.E. (1974). Seeding habits of upper-slope tree species. III. Dispersal of white and shasta red fir seeds on a clear-cut. *US Department of Agriculture, Forest Service, Research Note PNW-215*.

Fryer, G.I. & Johnson, E.A. (1988). Reconstructing fire behaviour and effects in a subalpine forest. *Journal of Applied Ecology* **25**: 1063–72.

Fuglem, P.L. & Murphy, P.J. (1980). Foliar moisture and crown fires in Alberta. *Alberta Energy and Natural Resources, ENR Report No. 158*. Edmonton.

Garcia, O. (1981). Simplified method-of-moments estimation for the Weibull distribution. *New Zealand Journal of Forestry Science* **11**: 304–6.

Gillespie, B.C. (1975). Territorial expansion of the Chipewyan in the 18th century. In *Proceedings: Northern Athapaskan Conference 1971, National Museums of Canada, National Museum of Man Mercury Series, Canadian Ethnology Service Paper No. 27*, ed. A. McFadyen Clark, vol. 2, pp. 350–88. Ottawa.

Green, L.R. (1981). Burning by prescription in chaparral. *US Department of Agriculture, Forest Service, General Technical Report PSW-51*.

Greene, D.F. (1989). The aerodynamics and dispersal of plumed and winged seeds. PhD thesis, University of Calgary, Canada.

Greene, D.F. & Johnson, E.A. (1989). A model of wind dispersal of winged or plumed seeds. *Ecology* **70**: 339–47.

Habeck, J.R. (1976). Forests, fuels and fire in the Selway–Bitterroot Wilderness, Idaho.

Proceedings of the Tall Timbers Fire Ecology Conference **14**: 305–53.

Hanes, T.L. (1971). Succession after fire in the chaparral of southern California. *Ecological Monographs* **41**: 27–52.

Hardy, C.E. & Franks, J.W. (1963). Forest fires in Alaska. *US Department of Agriculture, Forest Service, Research Paper INT-5.*

Hardy, Y.J., Lafond, A. & Hamel, L. (1983). The epidemiology of the current spruce budworm outbreak in Québec. *Forest Science* **29**: 715–25.

Hardy, Y.J., Mainville, M. & Schmitt, D.M. (1985). An atlas of spruce budworm defoliation in eastern North America, 1938–80. *US Department of Agriculture, Forest Service, Miscellaneous Publication No. 1449.*

Hare, F.K. & Hay, J.E. (1974). The climate of Canada and Alaska. In *World Survey of Climatology. Volume 11: Climates of North America,* ed. R.A. Bryson & F.K. Hare, pp. 49–192. Amsterdam: Elsevier.

Hare, F.K. & Ritchie, J.C. (1972). The boreal bioclimates. *Geographical Review* **62**: 333–65.

Hare, R.C. (1965). Contribution of bark to fire resistance of southern trees. *Journal of Forestry* **63**: 248–51.

Harrington, J.B. (1982). A statistical study of area burned by wildfire in Canada 1953–1980. *Environment Canada, Canadian Forestry Service, Petawawa National Forestry Institute, Information Report PI-X-16.*

Harrington, J. & Flannigan, M. (1987). Drought persistence at forested Canadian stations. *Preprint Volume, Ninth Conference on Fire and Forest Meteorology, April 21–24, 1987, San Diego, California,* pp. 204–6. Boston: American Meteorological Society.

Harter, H.L. & Moore, A.H. (1965). Maximum-likelihood estimation of the parameters of gamma and Weibull populations from complete and from censored samples. *Technometrics* **7**: 639–43.

Harvey, D.A. (1979). Lightning means in Alberta 1972–77. *Alberta Energy and Natural Resources, Alberta Forest Service, ENR Report 138.* Edmonton, Alberta.

Heidenrich, C.E. & Ray, A.J. (1976). *The Early Fur Trades: a Study in Cultural Interaction.* Toronto: McClelland & Stewart Ltd.

Heinselman, M.L. (1973). Fire in the virgin forests of the Boundary Waters Canoe Area, Minnesota. *Quaternary Research* **3**: 329–82.

Hellum, A.K. & Pelchat, M. (1979). Temperature and time affect the release and quality of seed from cones of lodgepole pine from Alberta. *Canadian Journal of Forest Research* **9**: 154–9.

Hemstrom, M.A. & Franklin, J.F. (1982). Fire and other disturbances of the forests in Mount Rainier National Park. *Quaternary Research* **18**: 32–51.

Hibbs, D.E. (1982). Gap dynamics in a hemlock-hardwood forest. *Canadian Journal of Forest Research* **12**: 522–7.

Hilado, C.J. (1977). The correlation of laboratory test results with behaviour in real fires. *Fire Flammability* **8**: 202–9.

Hottel, H.C., Williams, G.C. & Steward, F.R. (1965). The modeling of firespread through a fuel bed. In *Tenth Symposium (International) on Combustion,* pp. 997–1007. Pittsburgh: The Combustion Institute.

Hutchinson, G.E. (1953). The concept of pattern in ecology. *Proceedings of the Academy of Natural Sciences (Philadelphia)* **105**: 1–12.

Jarvenpa, R. (1980). The trappers of Patuanak: toward a spatial ecology of modern hunters. *National Museums of Canada, National Museum of Man Mercury Series, Canadian Ethnology Service Paper No. 67.* Ottawa.

Johnson, E.A. (1975). Buried seed populations in the subarctic forest east of Great Slave

Lake, Northwest Territories. *Canadian Journal of Botany* **53**: 2933–41.

Johnson, E.A. (1979). Fire recurrence in the subarctic and its implications for vegetation composition. *Canadian Journal of Botany* **57**: 1374–9.

Johnson, E.A. (1981). Vegetation organization and dynamics of lichen woodland communities in the Northwest Territories, Canada. *Ecology* **62**: 200–15.

Johnson, E.A. (1985). Disturbance: the process and the response. An epilogue. *Canadian Journal of Forest Research* **15**: 292–3.

Johnson, E.A. & Fryer, G.I. (1987). Historical vegetation change in the Kananaskis Valley, Canadian Rockies. *Canadian Journal of Botany* **65**: 853–8.

Johnson, E.A. & Fryer, G.I. (1989). Population dynamics in lodgepole pine–Engelmann spruce forests. *Ecology* **70**: 1335–45.

Johnson, E.A. Fryer, G.I. & Heathcott, M.J. (1990). The influence of man and climate on frequency of fire in the interior wet belt forest, British Columbia. *Journal of Ecology* **78**: 403–12.

Johnson, E.A. & Rowe, J.S. (1975). Fire in the subarctic wintering ground of the Beverley caribou herd. *American Midland Naturalist* **94**: 1–14.

Johnson, E.A. & Van Wagner, C.E. (1985). The theory and use of two fire history models. *Canadian Journal of Forest Research* **15**: 214–20.

Johnston, M. & Woodard, P. (1985). The effect of fire severity level on postfire recovery of hazel and raspberry in east-central Alberta. *Canadian Journal of Botony* **63**: 672–7.

Kao, J.H.K. (1959). A graphical estimation of mixed Weibull parameters in life-testing of electron tubes. *Technometrics* **1**: 389–407.

Kayll, A.J. (1963). Heat tolerance of scots pine seedling cambium using tetrazolium chloride to test viability. *Canada Department of Forestry Publication 1006*.

Kayll, A.J. (1968). Heat tolerance of tree seedlings. *Proceedings of the Tall Timbers Fire Ecology Conference* **8**: 89–105.

Kellman, M.C. (1970). The viable seed content of some forest soil in coastal British Columbia. *Canadian Journal of Botany* **48**: 1383–5.

Kendall, G.R. & Petrie, A.G. (1962). The frequency of thunderstorm days in Canada. *Meteorological Branch, Department of Transport, Canada. Toronto, Ontario, CIR-3688, TEC-418.*

Kiil, A.D. (1975). Fire spread in a black spruce stand. *Canadian Forestry Service, Bi-Monthly Research Notes* **31**: 2–3.

Kiil, A.D. & Grigel J.E. (1969). The May 1968 forest conflagrations in central Alberta – a review of fire weather, fuels and fire behavior. *Canadian Forestry Service, Northern Forest Research Centre Information Report A-X-24.*

King, J.R. (1971). *Probability charts for decision making.* New York: Industrial Press.

Knight, D.H. (1987). Parasites, lightning, and the vegetation mosaic in wilderness landscapes. In *Landscape Heterogeneity and Disturbance*, ed. M.G. Turner, pp. 59–83. New York: Springer-Verlag.

Knox, J.L. & Lawford, R.G. (1990). The relationship between Canadian prairie dry and wet months and circulation anomalies in the mid-troposphere. *Atmosphere–Ocean* **28**: 189–215.

Kramer, N.B. & Johnson, F.D. (1987). Mature forest seed banks of three habitat types in central Idaho. *Canadian Journal of Botany* **65**: 1961–6.

Kreith, F. (1965). *Principles of Heat Transfer*, 2nd edn. Scranton, Pennsylvania: International Textbook Company.

Krider, E.P., Noggle, R.C., Pifer, A.E. & Vance, D.L. (1980). Lightning direction-finding systems for forest fire detection. *Bulletin of the American Meteorological Society* **61**: 980–6.

Lanoville, R.A. & Schmidt, R.E. (1985). Wildfire documentation in the Northwest Territories: a case study of Fort Simpson Fire 40, 1983. *Proceedings Second Western Region Fire Weather Committee Science and Technology Seminar*, pp. 17–22. Canadian Forestry Service, Study NOR-5-191.

Larsen, C.P.S. (1989). Fine resolution palaeoecology in the boreal forest of Alberta: a long-term record of fire-vegetation dynamics. M.Sc. Thesis, McMaster University, Canada.

Lawson, B.D. (1973). Fire behavior in lodgepole pine stands related to the Canadian Fire Weather Index. *Department of the Environment, Canadian Forestry Service, Information Report BC-X-76.*

Lawson, B.D., Stocks, B.J., Alexander, M.E. & Van Wagner, C.E. (1985). A system for predicting fire behavior in Canadian forests. In *Proceedings of the Eighth Conference on Fire and Forest Meteorology, April 29–May 2, Detroit, Mich.*, pp. 6–16, Society of American Foresters, Bethesda, MD: SAF Publication 85–04.

Légère, A. & Payette, S. (1981). Ecology of a black spruce (*Picea mariana*) clonal population in the Hemiarctic Zone, northern Quebec: population dynamics and spatial development. *Arctic and Alpine Research* **13**: 261–76.

Lewis, H.T. (1977). Maskuta: the ecology of Indian fires in northern Alberta. *Western Canadian Journal of Anthropology* **7**: 15–52.

Lewis, H.T. (1982). *A time for burning. Occasional Publication Number 17.* Boreal Institute for Northern Studies, University of Alberta.

Little, C.H.A. (1970). Seasonal changes in carbohydrate and moisture content in needles of balsam fir (*Abies balsamea*). *Canadian Journal of Botony* **48**: 2021–8.

Lorimer, C.G. (1985). Methodological considerations in the analysis of forest disturbance history. *Canadian Journal of Forest Research* **15**: 200–13.

Madany, M.H., Swetnam, T.W. & West, N.E. (1982). Comparison of two approaches for determining fire dates from tree scars. *Forest Science* **28**: 856–61.

Mak, E.H.T. (1988). Measuring foliar flammability with the limiting oxygen index method. *Forest Science* **34**: 523–9.

Martin, R.E. (1963). Thermal properties of bark. *Forest Products Journal* **13**: 419–26.

Masters, A.M. (1990). Changes in forest fire frequency in Kootenay National Park, Canadian Rockies. *Canadian Journal of Botony* **68**: 1763–7.

May, R.M. (1973). *Stability and Complexity in Model Ecosystems.* Princeton, NJ: Princeton University Press.

McBride, J.R. (1983). Analysis of tree rings and fire scars to establish fire history. *Tree-Ring Bulletin* **43**: 51–67.

McConkey, T.W. & Gedney, D.R. (1951). A guide for salvaging white pine injured by forest fires. *US Department of Agriculture, Forest Service, Research Note RM-11.*

McLean, A. (1969). Fire resistance of forest species as influenced by root systems. *Journal of Range Management* **22**: 120–2.

Merrill, D.F. & Alexander, M.E. (ed.) (1987). *Glossary of forest fire management terms*, 4th edn. National Research Council of Canada, Canadian Committee Forest Fire Management, Ottawa, Ontario. Publication NRCC26516.

Methven, I.R. (1971). Prescribed fire, crown scorch and mortality: field and laboratory studies on red and white pine. *Canadian Forestry Service Information Report PS-X-31.*

Minnich, R.A. (1983). Fire mosaics in Southern California and Northern Baja California. *Science* **219**: 1287–94.

Minnich, R.A. (1987). Fire behavior in Southern California chaparral before fire control: the Mount Wilson burns at the turn of the century. *Annals of the Association of American Geographers* **77**: 599–618.

Mohler, C.L., Marks, P.L. & Sprugel, D.G. (1978). Stand structure and allometry of trees during self-thinning of pure stands. *Journal of Ecology* **66**: 599–614.

Moore, J.M. & Wein, R.W. (1977). Viable seed populations by soil depth and potential site recolonization after disturbance. *Canadian Journal of Botany* **55**: 2408–12.

Morin, H. (1990). Analyse dendroécologique d'une sapinière issue d'un chablis dans la zone boréale, Québec. *Canadian Journal of Forest Research* **20**: 1753–8.

Morin, H. & Laprise, D. (1990). Histoire récente des épidémies de la Tordeuse des bourgeons de l'épinette au nord du lac Saint-Jean (Québec: une analyse dendrochronologique). *Canadian Journal of Forest Research* **20**: 1–8.

Murphy, P.J. & Tymstra, C. (1986). The 1950 Chinchaga River fire in the Peace River region of British Columbia/Alberta: preliminary results of simulating forward spread distances. In *Proceedings of the Third Western Region Fire Weather Committee Scientific and Technical Seminar*, Feb. 4, 1986, Edmonton, ed. M.E. Alexander, pp. 20–30. Canadian Forestry Service, Northern Forestry Centre, Edmonton, Alberta. Study NOR-5-05.

Mutch, R.W. & Gastineau, O.W. (1970). Timelag and equilibrium moisture content of reindeer lichen. *US Department of Agriculture, Forest Service, Research Paper INT-76*.

Nageli, W. (1953). Die Windbremsung durch einen grösseren Waldcomplex: ein Bertrag zum Problem der Schutzstreifenbreite bei Windshutzan lagen. *Compt. Rend II*, pp. 240–6. International Union of Forest Research Organizations, Firenze.

Nelson, R.M., Jr (1969). Some factors affecting the moisture timelags of woody materials. *US Department of Agriculture, Forest Service, Research Paper SE-44*.

Nelson, R.M., Jr (1980). Flame characteristics for fires in southern fuels. *US Department of Agriculture, Forest Service, Research Paper SE-205*.

Nelson, R.M., Jr & Adkins, C.W. (1986). Flame characteristics of wind-driven surface fires. *Canadian Journal of Forest Research* **16**: 1293–300.

Newark, M.J. (1975). The relationship between forest fire occurrence and 500 mb longwave ridging. *Atmosphere* **13**: 26–33.

Niessen, W. von & Blumen, A. (1986). Dynamics of forest fires as a directed percolation model. *Journal of Physics A: Mathematical and General* **19**: L289–93.

Niessen, W. von & Blumen, A. (1988). Dynamic simulation of forest fires. *Canadian Journal of Forest Research* **18**: 805–12.

Nimchuk, N. (1983). *Wildfire behavior associated with upper ridge breakdown.* Alberta Energy and Natural Resources, Forest Service, Edmonton, Alberta. ENR Report Number T/50.

Norum, R.A. (1977). Preliminary guidelines for prescribed burning under standing timber in western larch/Douglas fir forests. *US Department of Agriculture, Forest Service, Research Note INT-229*.

Noste, N.V. & Bushey, C.L. (1987). Fire response of shrubs of dry forest habitat types in Montana and Idaho. *US Department of Agriculture, Forest Service, General Technical Report INT-239*.

Ohtsuki, T. & Keyes, T. (1986). Biased percolation: forest fires with wind. *Journal of Physics A: Mathematical and General* **19**: L281–7.

Oliver, C.D. (1981). Forest development in North America following major disturbances. *Forest Ecology and Management* **3**: 153–68.

Pagni, P.J. & Peterson, T.G. (1972). Flame spread through porous fuels. *Fourteenth Symposium (International) on Combustion*, pp. 1099–107. Pittsburgh: The Combustion Institute.

Parsons, D.J. (1976). The role of fire in natural communities: an example from the southern Sierra Nevada, California. *Environmental Conservation* **3**: 91–9.

Patterson, W.A. III, Edwards, K.J. & Maguire, D.J. (1987). Microscopic charcoal as a fossil indicator of fire. *Quaternary Science Reviews* 6: 3–23.

Payette, S. & Filion, L. (1985). White spruce expansion at the tree line and recent climatic change. *Canadian Journal of Forest Research* 15: 241–51.

Payette, S., Filion, L., Gauthier, L. & Boutin, Y. (1985). Secular climate change in old-growth tree-line vegetation of northern Quebec. *Nature* 315: 135–8.

Payette, S. & Gagnon, R. (1979). Tree-line dynamics in Ungava peninsula, Northern Quebec. *Holarctic Ecology* 2: 239–48.

Payette, S., Morneau, C., Sirois, L. & Desponts, M. (1989). Recent fire history of the northern Quebec biomes. *Ecology* 70: 656–73.

Perry, J.H. (ed.) (1963). *Chemical Engineers' Handbook*. New York: McGraw-Hill.

Peterson, D.L. (1983). Predicting fire-caused mortality in four northern Rocky Mountain conifers. In *New Forests for a Changing World: Proceedings of the 1983 Convention of the Society of American Foresters*, pp. 276–80, Bethesda, Maryland.

Peterson, D.L. & Ryan, K.C. (1986). Modeling postfire conifer mortality for long-range planning. *Environmental Management* 10: 797–808.

Petty, J.A. & Swain, C. (1985). Factors influencing stem breakage of conifers in high winds. *Forestry* 58: 75–84.

Pharis, R.P. (1967). Seasonal fluctuation in the foliage-moisture content of well-watered conifers. *Botanical Gazette* 128: 179–85.

Purdie, R.W. (1977a). Early stages of regeneration after burning in dry Sclerophyll vegetation. I. Regeneration of the understorey by vegetative means. *Australian Journal of Botany* 25: 21–34.

Purdie, R.W. (1977b). Early stages of regeneration after burning in dry Sclerophyll vegetation. II. Regeneration by seed germination. *Australian Journal of Botany* 25: 35–46.

Quintilio, D., Fahnestock, G.R. & Dubé, D.E. (1977). Fire behavior in upland jack pine: the Darwin Lake Project. *Environment Canada, Canadian Forestry Service, Northern Forest Research Centre, Information Report NOR-X-174.*

Ray, A.J. (1974). *Indians in the fur trade: their role as trappers, hunters, and middlemen in the lands southwest of Hudson Bay, 1660–1870.* Toronto: University of Toronto Press.

Requa, L.E. (1964). Lightning behavior in the Yukon. *Proceedings, Tall Timber Fire Ecology Conference* 3: 111–19.

Riggan, P.J., Goode, S., Jacks, P.M. & Lockwood, R.N. (1988). Interaction of fire and community development in chaparral of southern California. *Ecological Monographs* 58: 155–76.

Ritchie, J.C. (1987). *Postglacial Vegetation of Canada*. Cambridge: Cambridge University Press.

Romme, W.H. (1982). Fire and landscape diversity in subalpine forests of Yellowstone National Park. *Ecological Monographs* 52: 199–221.

Romme, W.H. & Knight, D.H. (1981). Fire frequency and subalpine forest succession along a topographic gradient in Wyoming. *Ecology* 62: 319–26.

Rothermel, R.C. (1972). A mathematical model for predicting fire spread in wildland fuels. *US Department of Agriculture, Forest Service, Research Paper INT-115.*

Rothermel, R.C. (1976). Forest fires and the chemistry of forest fuels. In *Thermal Uses and Properties of Carbohydrates and Lignins*, ed. F. Shafizadeh, K.V. Sarkanen & D.A. Tillman, pp. 245–59. New York: Academic Press.

Rothermel, R.C. (1983). How to predict the spread and intensity of forest and range fires. *US Department of Agriculture, Forest Service, General Technical Report INT-143.*

Rothermel, R.C. & Anderson, H.E. (1966). Fire spread characteristics determined in the laboratory. *US Department of Agriculture, Forest Service, Research Paper INT-30.*

Rowe, J.S. (1961). Critique of some vegetational concepts as applied to forests of northwestern Alberta. *Canadian Journal of Botany* **39**: 1007–17.

Rowe, J.S. (1970). Spruce and fire in northwest Canada and Alaska. *Proceedings of the Tall Timbers Fire Ecology Conference* **10**: 245–54.

Runkle, J.R. (1985). Disturbance regimes in temperate forests. In *The Ecology of Natural Disturbance and Patch Dynamics*, ed. S.T.A. Pickett & P.S. White, pp. 17–33. New York: Academic Press.

Ryan, K.C. (1982). Evaluating potential tree mortality from prescribed burning. In *Proceedings of the Symposium on Site Preparation and Fuels Management on Steep Terrain*, ed. D.M. Baumgartner, pp. 167–79. Washington State University Cooperative Extension, Pullman, Washington.

Sandberg, D.V. (1980). Duff reduction by prescribed underburning in Douglas fir. *US Department of Agriculture, Forest Service, Research Paper PNW-272.*

Sando, R.W. & Haines, D.A. (1972). Fire weather and behavior of the Little Sioux Fire. *US Department of Agriculture, Forest Service, Research Report NC-76.*

Schaetzl, R.J., Burns, S.F., Johnson, D.L. & Small, T.W. (1989). Tree uprooting: review of impact on forest ecology. *Vegetatio* **79**: 165–76.

Schlesinger, W.H. & Gill, D.S. (1978). Demographic studies of the chaparral shrub, *Ceanothus megacarpus*, in the Santa Ynez Mountains, California. *Ecology* **59**: 1256–63.

Schroeder, M.J. (1950). The Hudson Bay high and the spring fire season in the lake states. *Bulletin of the American Meteorological Society* **31**: 111–18.

Schroeder, M.J., Glovinsky, M. *et al.* (1964). *Synoptic weather types associated with critical fire weather.* US Department of Agriculture, Forest Service, Pacific Southwest Forest and Range Experiment Station, Berkeley, California.

Shafizadeh, F. (1968). Pyrolysis and combustion of cellulosic materials. *Advances in Carbohydrate Chemistry* **23**: 419–74.

Shafizadeh, F. & DeGroot, W.F. (1976). Combustion characteristics of cellulosic fuels. In *Thermal Uses and Properties of Carbohydrates and Lignins*, ed. F. Shafizadeh, K.V. Sarkanen, & D.A. Tillman, pp. 1–17. New York: Academic Press.

Shearer, R.C. (1975). Seedbed characteristics in western larch forests after prescribed burning. *US Department of Agriculture, Forest Service, INT-167.*

Sheshukov, M.A. (1970). Effect of steepness of slope on the propagation rate of fire. *Lesnoe Khozyaistvo*, 1970: 50–4. (Translation No. 185672, Secretary of State, Ottawa.)

Shugart, H.H. (1984). *A Theory of Forest Dynamics.* New York: Springer-Verlag.

Siggins, H. (1933). Distribution and rate of fall of conifer seeds. *Journal of Agricultural Research* **47**: 119–28.

Simard, A.J. (1968). The moisture content of forest fuels. I. A review of the basic concepts. *Canada Department of Forestry and Rural Development, Forest Fire Research Institute, Information Report FF-X-14.*

Simard, A.J. (1975). *Forest Fire Weather Zones of Canada.* Environment Canada, Canada Forestry Service, Ottawa, Ontario (Map).

Sirois, L. & Payette, S. (1989). Postfire black spruce establishment in subarctic and boreal Quebec. *Canadian Journal of Forest Research* **19**: 1571–80.

Smith, W.L. & Leadbetter, M.R. (1963). On the renewal function for the Weibull distribution. *Technometrics* **5**: 393–6.

Sneeuwjagt, R.J. & Frandsen, W.H. (1977). Behavior of experimental grass fires vs. predictions based on Rothermel's fire model. *Canadian Journal of Forest Research* **7**: 357–67.

Springer, E.A. & Van Wagner, C.E. (1984). The seasonal foliar moisture trend of black spruce at Kapuskasing, Ontario. *Canadian Forestry Service Bi-monthly Research Notes* **4**: 39–42.

Sprugel, D.G. (1976). Dynamic structure of wave-regenerated *Abies balsamea* forests in the northeastern United States. *Journal of Ecology* **64**: 889–911.

Sprugel, D.G. (1989). The relationship of evergreenness, crown architecture, and leaf size. *American Naturalist* **133**: 465–79.

Stephenson, N.L. (1987). Use of tree aggregations in forest ecology and management. *Environmental Management* **11**: 1–5.

Stocks, B.J. (1970). Moisture in the forest floor – its distribution and movement. *Department of Fisheries and Forestry, Canadian Forestry Service Publication Number 1271.*

Stocks, B.J. (1974). Wildfires and the fire weather index system in Ontario. Department of the Environment, *Canadian Forestry Service, Great Lakes Forest Research Centre, Information Report O-X-213.*

Stocks, B.J. (1983). The 1980 forest fire season: its impact in west-central Canada. In *Preprint volume: Seventh Conference on Fire and Forest Meteorology, April 25–28, 1983, Ft Collins, Colorado.* Boston, MA: American Meteorological Society.

Stocks, B.J. (1987a). Fire behavior in immature jack pine. *Canadian Journal of Forest Research* **17**: 80–6.

Stocks, B.J. (1987b). Fire potential in the spruce budworm-damaged forests of Ontario. *Forestry Chronicle* **63**: 8–14.

Stocks, B.J. & Hartley, G.R. (1979). *Forest fire occurrence in Ontario.* Environment Canada, Canada Forestry Service, Great Lakes Forest Research Centre, Sault Ste. Marie, Ontario (Map).

Stocks, B.J. & Street, R.B. (1983). Forest fire weather and wildfire occurrence in the boreal forest of northwestern Ontario. In *Resources and Dynamics of the Boreal Zone*, ed. R.W. Wein, R.R. Riewe and I.R. Methven, pp. 249–65. Ottawa, Ontario: Association of Canadian Universities Northern Studies.

Stocks, B.J. & Walker, J.D. (1973). Climatic conditions before and during four significant forest fire situations in Ontario. *Canada Forest Service, Information Report O-X-187.*

Strickler, G.S. & Edgerton, P.J. (1976). Emergent seedlings from coniferous litter and soil in eastern Oregon. *Ecology* **57**: 801–7.

Suffling, R., Smith, B. & Dal Molin, J. (1982). Estimating past forest age distributions and disturbance rates in northwestern Ontario: a demographic approach. *Journal of Environmental Management* **14**: 45–56.

Susott, R.A. (1982). Characterization of the thermal properties of forest fuels by combustible gas analysis. *Forest Science* **28**: 404–20.

Susott, R.A., DeGroot, W.F. & Shafizadeh, F. (1975). Heat content of natural fuels. *Journal of Fire and Flammability* **6**: 311–25.

Swain, A.M. (1973). A history of fire and vegetation in northeastern Minnesota as recorded in lake sediments. *Quaternary Research*, **3**: 383–96.

Swain, A.M. (1980). Landscape patterns and forest history in the Boundary Waters Canoe Area, Minnesota: a pollen study from Hug Lake. *Ecology* **61**: 747–54.

Sylvester, T.W. & Wein, R.W. (1981). Fuel characteristics of arctic plant species and simulated plant community flammability by Rothermel's model. *Canadian Journal of Botany* **59**: 898–907.

Tande, G.F. (1979). Fire history and vegetation pattern of coniferous forests in Jasper National Park, Alberta. *Canadian Journal of Botany* **57**: 1912–31.

Tangren, C.D. (1976). The trouble with fire intensity. *Fire Technology* **12**: 261–5.

Thomas, P.A. & Wein, R.W. (1985a). The influence of shelter and the hypothetical effect of fire severity on the postfire establishment of conifers from seed. *Canadian Journal of Forest Research* **15**: 148–55.

Thomas, P.A. & Wein, R.W. (1985b). Delayed emergence of four conifer species on postfire seedbeds in eastern Canada. *Canadian Journal of Forest Research* **15**: 727–9.

Thomas, P.H. (1963). The size of flames from natural fires. *Ninth Symposium (International) on Combustion*, pp. 844–59. Pittsburgh: the Combustion Institute.

Thomas, P.H. (1967). Some aspects of the growth and spread of fire in the open. *Forestry* **40**: 139–64.

Treidl, R.A., Birch, E.C. & Sajecki, P. (1981). Blocking action in the northern hemisphere: a climatological study. *Atmosphere–Ocean* **19**: 1–23.

Van Cleve, K., Barney, R. & Schlentner, R. (1981). Evidence of temperature control of production and nutrient cycling in two interior Alaska black spruce ecosystems. *Canadian Journal of Forest Research* **11**: 258–73.

van der Pijl, L. (1972). *Principles of dispersal in higher plants*, 2nd edn. New York: Springer-Verlag.

Van Wagner, C.E. (1965). Describing forest fires – old ways and new. *Forestry Chronicle* **41**: 301–5.

Van Wagner, C.E. (1967). Seasonal variation in moisture content of eastern Canadian tree foliage and the possible effect on crown fire. *Canada Department of Forestry and Rural Development, Forestry Branch, Departmental Publication No. 1204.*

Van Wagner, C.E. (1968). Fire behavior mechanisms in a red pine plantation: field and laboratory evidence. *Canada Department of Forestry and Rural Development, Departmental Publication No. 1229.*

Van Wagner, C.E. (1969a). Drying rates of some fine forest fuels. *Fire Control Notes* **30**: 5, 7, and 12.

Van Wagner, C.E. (1969b). A simple fire-growth model. *Forestry Chronicle* **45**: 103–4.

Van Wagner, C.E. (1970). An index to estimate the current moisture content of the forest floor. Department of Fisheries and Forestry, Canadian Forest Service, Publication No. 1288.

Van Wagner, C.E. (1971a). Two solitudes in forest fire research. *Canadian Forestry Service Information Report PS-X-29.*

Van Wagner, C.E. (1971b). Fire and red pine. *Proceedings of the Tall Timbers Fire Ecology Conference* **10**: 211–19.

Van Wagner, C.E. (1972a). Duff consumption by fire in eastern pine stands. *Canadian Journal of Forest Research* **2**: 34–9.

Van Wagner, C.E. (1972b). Heat of combustion, heat yield, and fire behaviour. *Environment Canada, Forestry Service, Information Report PS-X-35.*

Van Wagner, C.E. (1973a). Height of crown scorch in forest fires. *Canadian Journal of Forest Research* **3**: 373–8.

Van Wagner, C.E. (1973b). Rough prediction of fire spread rates by fuel type. *Environment Canada, Canadian Forestry Service, Information Report PS-X-42.*

Van Wagner, C.E. (1974). A spread index for crown fires in spring. *Environment Canada, Forestry Service, Information Report PS-X-55.*

Van Wagner, C.E. (1975). Convection temperatures above low intensity forest fires. *Canadian Forestry Service, Bi-monthly Research Notes* **31**: 21.

Van Wagner, C.E. (1977a). Conditions for the start and spread of crown fire. *Canadian Journal of Forest Research* **7**: 23–34.

Van Wagner, C.E. (1977b). Effect of slope on fire spread rate. *Canadian Forestry Service, Bi-monthly Research Notes* **33**(1): 7–8.

Van Wagner, C.E. (1977c). Reader's Forum. *Fire Technology* **13**: 349–50.

Van Wagner, C.E. (1978). Age-class distribution and the forest fire cycle. *Canadian Journal of Forest Research* **8**: 220–7.

Van Wagner, C.E. (1979). A laboratory study of weather effects on the drying rate of jack pine litter. *Canadian Journal of Forest Research* **9**: 267–75.

Van Wagner, C.E. (1982). Initial moisture content and the exponential drying process. *Canadian Journal of Forest Research* **12**: 90–2.

Van Wagner, C.E. (1983). Fire behavior in northern conifer forests and shrublands. In *The Role of Northern Circumpolar Ecosystems*, ed. R.W. Wein & D.A. MacLean, pp. 65–80. New York: John Wiley & Sons.

Van Wagner, C.E. (1987). Development and structure of the Canadian Forest Fire Weather Index System. *Canadian Forestry Service, Forest Technical Report 35*.

Van Wagner, C.E. (1988a). The historical pattern of annual burned area in Canada. *Forestry Chronicle* **64**: 182–5. Erratum *64*:319.

Van Wagner, C.E. (1988b). Effect of slope on fires spreading downhill. *Canadian Journal of Forest Research* **18**: 818–20.

Van Wagner, C.E. & Methven, I.R. (1978). Discussion: Two recent articles on fire ecology. *Canadian Journal of Forest Research* **8**: 491–2.

Vincent, A.B. (1965). Black spruce – A review of its silvics, ecology and silviculture. *Department of Forestry, Canada, Publication Number 1100*.

Vogt, K.A., Grier, C.C. & Vogt, D.J. (1986). Production, turnover, and nutrient dynamics of above- and belowground detritus of world forests. *Advances in Ecological Research* **15**: 303–77.

Waldron, R.M. (1965). Cone production and seedfall in a mature white spruce stand. *Forestry Chronicle* **41**: 316–29.

Weber, M.G., Hummel, M. & Van Wagner, C.E. (1987). Selected parameters of fire behavior and *Pinus banksiana* Lamb. regeneration in eastern Ontario. *Forestry Chronicle* **63**: 340–6.

Wein, R.W. (1976). Frequency and characteristics of Arctic Tundra fires. *Arctic* **29**: 213–22.

Wendland, W.M. & Bryson, R.A. (1981). Northern hemisphere airstream regions. *Monthly Weather Review* **109**: 255–70.

Whipple, S.A. & Dix, R.L. (1979). Age structure and successional dynamics of a Colorado subalpine forest. *American Midland Naturalist* **101**: 142–58.

Wright, J.G. (1932). Forest-fire hazard research as developed and conducted at the Petawawa Forest Experiment Station. Canada Department of Forestry and Rural Development, Forestry Branch. Reprinted 1967. Forest Fire Research Institute, Ottawa, Ontario. Information Report FF-X-5.

Yarie, J. (1981). Forest fire cycles and life tables: a case study from interior Alaska. *Canadian Journal of Forest research* **11**: 554–62.

Yarranton, M. & Yarranton, G.A. (1975). Demography of a jack pine stand. *Canadian Journal of Botany* **53**: 310–14.

Yerbury, J.C. (1986). *The subarctic Indians and the fur trade, 1680–1860*. Vancouver: University of British Columbia Press.

Zackrisson, O. (1977). Influence of forest fires on the North Swedish boreal forest. *Oikos* **29**: 22–32.

Zasada, J.C., Norum, R.A., Van Veldhuizen, R.M. & Teutsch, C.E. (1983). Artificial regeneration of trees and tall shrubs in experimentally burned upland black spruce/feather moss stands in Alaska. *Canadian Journal of Forest Research* **13**: 903–13.

Zasada, J.C., Norum, R.A., Teutsch, C.E. & Densmore, R. (1987). Survival and growth of planted black spruce, alder, aspen and willow after fire on black spruce/feather moss sites in interior Alaska. *Forestry Chronicle* **63**: 84–8.

Index

127